FUNCTIONAL ANALYSIS and APPROXIMATION THEORY in NUMERICAL ANALYSIS

R. S. VARGA
Kent State University

SOCIETY for INDUSTRIAL and APPLIED MATHEMATICS

PHILADELPHIA, PENNSYLVANIA 19103

Copyright 1971 by the Society for Industrial and Applied Mathematics.
Second printing 1979.
Third printing 1984.
Fourth printing 1991.

Printed by J. W. Arrowsmith Ltd., Bristol BS3 2NT, England.

FUNCTIONAL ANALYSIS AND APPROXIMATION THEORY IN NUMERICAL ANALYSIS

Contents

This volume is dedicated to
GARRETT BIRKHOFF
on the occasion of his sixtieth birthday

Preface

The purpose of these lecture notes is to survey in part the enormously expanding literature on the numerical approximation of solutions of elliptic boundary value problems by means of variational and finite element methods. Surveying this area will, as we shall see, require almost constant application of results and techniques from functional analysis and approximation theory to the field of numerical analysis, and it is our hope that the material presented here will serve to stimulate further activity which will strengthen the ties already connecting these fields.

Although our primary interest will concern the numerical approximation of elliptic boundary value problems, the methods to be described lend themselves as well rather naturally to discussions concerning eigenvalue problems and initial value problems, such as the heat equation. On the negative side, it is unfortunate that almost nothing will be said here about *scientific computing*, i.e., the real problems of implementation of such mathematical theories to working programs on high-speed computers, and the numerical experience which has already been gained on such problems. Fortunately, scientific computing is one of the key points of the monograph by Professor Garrett Birkhoff,[1] and we are grateful to be able to refer the reader to this work.

The intent of these lecture notes is to make each portion of the notes roughly independent of the remaining material. This is why the references used in each of the nine chapters are compiled separately at the end of each chapter.

It is a sincere pleasure to acknowledge the support of the National Science Foundation under a grant to the Conference Board of the Mathematical Sciences, for the Regional Conference held at Boston University July 20–24, 1970, and to acknowledge Professor Robin Esch's superb handling of even the most minute details of this Conference in Boston. Without his untiring efforts the Conference would not have been a success.

It is also a pleasure to acknowledge the fact that these notes benefited greatly from suggestions and comments by Garrett Birkhoff, James Dailey, George Fix, John Pierce and Blair Swartz. Finally, we thank Mrs. Julia Froble for her careful typing of the manuscript.

RICHARD S. VARGA

---

[1] Garrett Birkhoff, *Numerical Solution of Elliptic Partial Differential Equations*, SIAM Publications, 1971, 78 pp.

CHAPTER 1

L-Splines

1.1. Basic theory. Splines, as is well known, were effectively introduced to the mathematical world by I. J. Schoenberg [1.1] in 1946, and splines have since become the focus of much mathematical activity. In particular, approximation theorists and numerical analysts have of late literally seized upon splines because of their many beautiful properties and because of their wide range of application to the numerical approximation of solutions of differential equations. *It is these beautiful properties and wide range of applications of splines which we propose to cover in part in these lectures.*

The mathematical development of the theory of splines since Schoenberg's fundamental paper in 1946 has been both extremely diverse and extremely rapid. Several recent books on splines (cf. Ahlberg, Nilson and Walsh [1.2], T. N. E. Greville [1.3], I. J. Schoenberg [1.4]) indeed attest to this rapid development. To describe the development of spline theory, we begin here with a study of L-*splines*. This is a somewhat middle ground in the development, in that the theory of L-splines is certainly not classical, nor is it the most general to date. However, as we shall see, most of what is obtained here for L-splines carries over to more general splines recently investigated by several authors.

To begin, for $-\infty < a < b < +\infty$ and for a positive integer N, let

$$(1.1.1) \qquad \Delta : a = x_0 < x_1 < \cdots < x_N = b$$

denote a partition of the interval $[a, b]$ with *knots* x_i. The collection of all such partitions Δ of $[a, b]$ is denoted by $\mathscr{P}(a, b)$. We further define

$$\bar{\pi} = \max_{0 \leq i \leq N-1} (x_{i+1} - x_i) \quad \text{and} \quad \underline{\pi} = \min_{0 \leq i \leq N-1} (x_{i+1} - x_i)$$

for each Δ of the form (1.1.1). For any $\sigma \geq 1$, $\mathscr{P}_\sigma(a, b)$ denotes the subset of all partitions in $\mathscr{P}(a, b)$ for which

$$(1.1.2) \qquad \bar{\pi}/\underline{\pi} \leq \sigma.$$

In particular, $\mathscr{P}_1(a, b)$ is the collection of all uniform partitions of $[a, b]$, and its elements are denoted by Δ_u.

For additonal notation, if $C^p[a, b]$ is the set of all real-valued functions which have continuous derivatives of order at least p in $[a, b]$, we then recall that the Sobolev space $W_q^s[a, b]$, where $1 \leq q \leq \infty$ and s is any nonnegative integer, is

This research was supported in part by AEC Grant (11-1)-2075.

defined as the completion of the set of all real-valued functions $f \in C^\infty[a, b]$ with respect to the norm:

$$\|f\|_{W_q^s[a,b]} \equiv \left\{ \sum_{j=0}^{s} \int_a^b |D^j f|^q \, dx \right\}^{1/q}, \qquad\qquad 1 \leqq q < \infty,$$

(1.1.3)

$$\|f\|_{W_\infty^s[a,b]} \equiv \sum_{j=0}^{s} \|D^j f\|_{L_\infty[a,b]} \left(= \sum_{j=0}^{s} \left\{ \max_{x \in [a,b]} |D^j f(x)| \right\} \right), \qquad q = \infty.$$

Equivalently, $W_q^s[a, b]$ is the collection of all real-valued functions f defined on $[a, b]$ with (for $s > 0$) $D^{s-1}f$ absolutely continuous on $[a, b]$ and $D^s f \in L_q[a, b]$. Clearly, for $s > 0$, $W_q^s[a, b] \subset C^{s-1}[a, b]$.

To describe L-splines, consider the linear differential operator L of order m:

$$\text{(1.1.4)} \qquad\qquad Lu(x) = \sum_{j=0}^{m} c_j(x) D^j u(x), \qquad m \geqq 1, \quad D^j \equiv \left(\frac{d}{dx} \right)^j,$$

where $c_j \in C^j[a, b], 0 \leqq j \leqq m$, with $c_m(x) \geqq \delta > 0$ for all $x \in [a, b]$. An important special case is the choice $L = D^m$. Next, let z be any (fixed) positive integer with $1 \leqq z \leqq m$. Then, $\text{Sp}(L, \Delta, z)$, the L-spline space, is the collection of all real-valued functions w defined on $[a, b]$ such that (cf. Ahlberg, Nilson and Walsh [1.2, Chap. 6], Greville [1.5], and Schultz and Varga [1.6])

$$L^*Lw(x) = 0, \quad x \in (a, b) - \{x_i\}_{i=1}^{N-1},$$

(1.1.5)

$$D^k w(x_i-) = D^k w(x_i+) \quad \text{for} \quad 0 \leqq k \leqq 2m - 1 - z, \quad 0 < i < N,$$

where L^* is the formal adjoint of L, i.e., $L^*v \equiv \sum_{j=0}^{m} (-1)^j D^j \{c_j(x)v(x)\}$. In other words, each $w \in \text{Sp}(L, \Delta, z)$ is locally a solution of $L^*Lw = 0$, pieced together at the interior knots x_i in such a way, depending on z, that $w \in C^{2m-z-1}[a, b]$. Thus, $\text{Sp}(L, \Delta, z) \subset C^{2m-z-1}[a, b]$, but because of the assumed smoothness of the coefficients c_j in (1.1.4), we can sharpen this inclusion to $\text{Sp}(L, \Delta, z) \subset W_\infty^{2m-z}[a, b]$. In addition, it can be verified that $\text{Sp}(L, \Delta, z)$ is a *linear* space of dimension $2m + z(N - 1)$.

In the important special case $L = D^m$, the elements of $\text{Sp}(D^m, \Delta, z)$ are, from (1.1.5), polynomials of degree $2m - 1$ on each subinterval of Δ, and as such are called *polynomial splines*. More specially, when $L = D^m$ and $z = m$, elements of the associated L-spline space are called *Hermite splines*, and the collection of such Hermite splines is denoted by $H^{(m)}(\Delta)$. From (1.1.5), $H^{(m)}(\Delta) \subset W_\infty^m[a, b] \subset C^{m-1}[a, b]$. Similarly, when $L = D^m$ and $z = 1$, the elements of the associated L-spline space are called simply *splines*, and the collection of such splines is denoted by $\text{Sp}^{(m)}(\Delta)$. From (1.1.5), $\text{Sp}^{(m)}(\Delta) \subset W_\infty^{2m-1}[a, b] \subset C^{2m-2}[a, b]$.

We now discuss the possibility of *interpolation* of given functions by elements in $\text{Sp}(L, \Delta, z)$. Given any $g \in C^{m-1}[a, b]$, it can be shown by elementary methods (cf. [1.6]) that there exists a *unique* $s \in \text{Sp}(L, \Delta, z)$ which interpolates g in the sense that

$$D^j(g - s)(x_i) = 0, \qquad\qquad 0 \leqq j \leqq z - 1, \quad 0 < i < N,$$

(1.1.6)

$$D^j(g - s)(a) = D^j(g - s)(b) = 0, \qquad\qquad 0 \leqq j \leqq m - 1.$$

In particular, since $W_2^m[a, b] \subset C^{m-1}[a, b]$, each element in $W_2^m[a, b]$ possesses a unique interpolant in $\mathrm{Sp}(L, \Delta, z)$, in the sense of (1.1.6). As an integration by parts shows, if $g \in W_2^m[a, b]$, and s is its interpolant of (1.1.6) in $\mathrm{Sp}(L, \Delta, z)$, the *first integral relation* (cf. [1.2, p. 205]) of

$$(1.1.7) \qquad \int_a^b (Lg)^2 \, dx = \int_a^b \{L(g - s)\}^2 \, dx + \int_a^b (Ls)^2 \, dx$$

is valid. Note, however, that s from (1.1.6) is necessarily also the unique interpolant in $\mathrm{Sp}(L, \Delta, z)$ of *any* $f \in W_2^m[a, b]$ for which

$$(1.1.8) \qquad \begin{aligned} D^j(g - f)(x_i) &= 0, & 0 \leqq j \leqq z - 1, \quad 0 < i < N, \\ D^j(g - f)(a) &= D^j(g - f)(b) = 0, & 0 \leqq j \leqq m - 1. \end{aligned}$$

Thus, the first integral relation is valid with g replaced by any such f:

$$\int_a^b (Lf)^2 \, dx = \int_a^b \{L(f - s)\}^2 \, dx + \int_a^b (Ls)^2 \, dx,$$

from which it follows that

$$\int_a^b (Lf)^2 \, dx \geqq \int_a^b (Ls)^2 \, dx.$$

The above inequality then has the following beautiful interpretation.

THEOREM 1.1. *Given any* $g \in W_2^m[a, b]$, *let* U_g *be the collection of all* $f \in W_2^m[a, b]$ *which satisfy* (1.1.8). *Then*

$$(1.1.9) \qquad \|Ls\|_{L_2[a,b]} = \inf_{f \in U_g} \|Lf\|_{L_2[a,b]},$$

where s *is the unique interpolant of* g *in* $\mathrm{Sp}(L, \Delta, z)$ *in the sense of* (1.1.6).

This first integral relation (1.1.7) is important in that it is the basis for the following error bounds of Theorem 1.2. Its proof is elementary, requiring, for example, in the case $q = +\infty$, just Rolle's theorem (cf. [1.6]).

THEOREM 1.2. *Given* $g \in W_2^m[a, b]$, *and given* $\Delta \in \mathscr{P}(a, b)$, *let* s *be the unique element in* $\mathrm{Sp}(L, \Delta, z)$ *which interpolates* g *in the sense of* (1.1.6). *Then, for* $2 \leqq q \leqq \infty$,

$$(1.1.10) \qquad \|D^j(g - s)\|_{L_q[a,b]} \leqq K \bar{\pi}^{m-j-1/2+1/q} \|g\|_{W_2^m[a,b]}, \quad 0 \leqq j \leqq m - 1.$$

For polynomial splines $(L = D^m)$, $\|g\|_{W_2^m[a,b]}$ *can be replaced by* $\|D^m g\|_{L_2[a,b]}$ *in* (1.1.10).

The constant K in (1.1.10) means here, as in the subsequent material, a generic constant which is independent of g, but is dependent on m, n, a, b, and σ if $\Delta \in \mathscr{P}_\sigma(a, b)$.

Several interesting remarks can be made about the error bounds of (1.1.10). Although (1.1.10) is established by elementary means, it is nevertheless the case that the exponent of $\bar{\pi}$ in (1.1.10) is *best possible* for the space $W_2^m[a, b]$, i.e., it cannot be increased for all $g \in W_2^m[a, b]$ (cf. Birkhoff, Schultz and Varga [1.7] and Golomb [1.8]).

It is also interesting to remark that the inequality of (1.1.10) can be shown to be *quasi-optimal* (cf. Babuška, Práger, Vitásek [1.9, p. 232]) in the sense of *n-widths* of Kolmogorov (cf. Lorentz [1.10, Chap. 9]), and such related ideas for splines have been studied by Aubin [1.11] and Golomb [1.12].

Error bounds analogous to Theorem 1.2 can be obtained for L-spline interpolation of smoother functions g. In particular, if $g \in W_2^{2m}[a, b]$ and s is its $\mathrm{Sp}(L, \Delta, z)$-interpolant in the sense of (1.1.6), an integration by parts again shows that the *second integral relation* (cf. [1.2, p. 205])

$$(1.1.11) \qquad \int_a^b \{L(g - s)\}^2 \, dx = \int_a^b (g - s)(L^*Lg) \, dx$$

is valid. This is similarly used in proving (cf. [1.13]) the error bounds of the following theorem.

THEOREM 1.3. *Given $g \in W_2^{2m}[a, b]$, and given $\Delta \in \mathscr{P}_\sigma(a, b)$, let s be the unique element in $\mathrm{Sp}(L, \Delta, z)$ which interpolates g in the sense of (1.1.6). Then, for $2 \leqq q \leqq \infty$,*

$$(1.1.12) \qquad \|D^j(g - s)\|_{L_q[a,b]} \leqq K\pi^{2m-j-1/2+1/q}\|g\|_{W_2^{2m}[a,b]}, \quad 0 \leqq j \leqq 2m - z.$$

For polynomial splines $(L = D^m)$, $\|g\|_{W_2^{2m}[a,b]}$ can be replaced by $\|D^{2m}g\|_{L_2[a,b]}$ in (1.1.12).

For $0 \leqq j \leqq m - 1$, it is worth noting that the error bounds of (1.1.12) are valid for *any* $\Delta \in \mathscr{P}(a, b)$. The exponent of π in (1.1.12) is again best possible for the space $W_2^{2m}[a, b]$ for general L-spline interpolation. However, in terms of error bounds for g in $W_\infty^{2m}[a, b]$ or $W_\infty^m[a, b]$ for polynomial spline interpolation, the exponent of π in (1.1.10) and (1.1.12) can in special cases be increased by $\frac{1}{2}$ when $q = +\infty$ (cf. Swartz and Varga [1.13]). Next, we also mention that the results of Theorems 1.1–1.3 are known to be valid for more general forms of boundary interpolation than considered in (1.1.6). In addition, it is also possible to *vary* the parameter z from knot to knot with *no* change in the interpolation error bounds. Such refinements can be found for example in [1.6] and [1.13].

From the interpolation error bounds of Theorems 1.2 and 1.3, one can deduce, via the use of *interpolation space theory* (to be described in § 1.2), analogous interpolation error bounds for functions g in spaces intermediate to $W_2^m[a, b]$ and $W_2^{2m}[a, b]$. But the desire is to obtain error bounds for functions g even *less* smooth than $C^{m-1}[a, b]$, and this poses a problem. Clearly, the interpolation of g, as defined in (1.1.6) needs the existence of derivatives of g of order $m - 1$ in $[a, b]$, and thus, a modification of the definition of interpolation in (1.1.6) is necessary. To do this, we make use of the familiar notion of *Lagrange polynomial interpolation*, as described in Davis [1.14, Chap. 2].

If $\Delta \in \mathscr{P}_\sigma(a, b)$ has at least $2m$ knots, let $\mathscr{L}_{2m-1,0}g$ denote the Lagrange polynomial interpolation (of degree $2m - 1$) of g in the knots $x_0, x_1, \cdots, x_{2m-1}$, i.e.,

$$(1.1.13) \qquad (\mathscr{L}_{2m-1,0}g)(x_j) = g(x_j), \qquad 0 \leqq j \leqq 2m - 1.$$

Then, although g need not possess derivatives through order $m - 1$ at $x = a$, $\mathscr{L}_{2m-1,0}g$ does, and we can define interpolation by an $s \in \mathrm{Sp}(L, \Delta, z)$ at $x = a$ now by means of

$$(1.1.14) \qquad D^j s(a) = D^j(\mathscr{L}_{2m-1,0}g)(a), \qquad 0 \leqq j \leqq m - 1.$$

Similar definitions of interpolation can be used at other knots of Δ. We now state a result of Swartz and Varga [1.13] (see also Schultz [1.15] for a related use of Lagrange polynomial interpolation).

THEOREM 1.4. *Given* $g \in C^k[a, b]$ *with* $0 \leq k < 2m$ *and given* $\Delta \in \mathscr{P}_\sigma(a, b)$ *with at least $2m$ knots, let s be the unique element in* $\mathrm{Sp}(L, \Delta, z)$ *which interpolates g in the sense that*

$$D^j\{s - \mathscr{L}_{2m-1,i}g\}(x_i) = 0, \qquad 0 \leqq j \leqq z - 1, \quad 0 < i < N,$$

$$(1.1.15)$$

$$D^j\{s - \mathscr{L}_{2m-1,0}g\}(a) = D^j\{s - \mathscr{L}_{2m-1,N}g\}(b) = 0, \ 0 \leqq j \leqq m - 1,$$

where $\mathscr{L}_{2m-1,i}g$ is the Lagrange polynomial interpolation of g in the $2m$ consecutive knots $x_{j_i}, x_{j_i+1}, \cdots, x_{j_i+2m-1}$, *where* $x_i \in [x_{j_i}, x_{j_i+2m-1}]$. *Then, for* $2 \leqq q \leqq \infty$,

$$K\pi^{k-j-1/2+1/q}\{\omega(D^k g, \pi) + \pi^{2m-k}\|g\|_{W_2^k[a,b]}\}$$

$$(1.1.16) \qquad \geqq \begin{cases} \|D^j(g - s)\|_{L_q[a,b]} & \text{if} \ \ 0 \leqq j \leqq \min(k, 2m - z), \\ \|D^j s\|_{L_q[a,b]} & \text{if} \ \ \min(k, 2m - z) < j \leqq 2m - z. \end{cases}$$

For polynomial splines $(L = D^m)$, *the term involving* $\|g\|_{W_2^k[a,b]}$ *in* (1.1.16) *can be deleted.*

In (1.1.16), we have used the notation

$$(1.1.17) \qquad \omega(f, h) = \sup\{|f(x + t) - f(x)| : x, x + t \text{ are in } [a, b] \text{ with } |t| \leqq h\}$$

to denote the usual *modulus of continuity* of any bounded function f defined on $[a, b]$.

For the extension of the result of Theorem 1.4 to Sobolev spaces, we have the following corollary (cf. [1.13]).

COROLLARY 1.5. *With the hypotheses of Theorem* 1.4, *if* $g \in W_r^{k+1}[a, b]$ *with* $1 \leqq r \leqq \infty$ *and* $0 \leqq k < 2m$, *then for* $\max(r, 2) \leqq q \leqq \infty$,

$$K\pi^{k+1-j+1/q+\min(-1/r,-1/2)}\|g\|_{W_r^{k+1}[a,b]}$$

$$(1.1.18) \qquad \geqq \begin{cases} \|D^j(g - s)\|_{L_q[a,b]} & \text{if} \ \ 0 \leqq j \leqq \min(k, 2m - z), \\ \|D^j s\|_{L_q[a,b]} & \text{if} \ \ \min(k, 2m - z) < j \leqq 2m - z. \end{cases}$$

For polynomial splines, $\|g\|_{W_r^{k+1}[a,b]}$ *can be replaced by* $\|D^{k+1}g\|_{L_r[a,b]}$ *in* (1.1.18).

We note that when $k = m - 1$ and $r = 2$, the first inequality of (1.1.18) reduces to the inequality of (1.1.10). Similarly, when $k = 2m - 1$ and $r = 2$, the first inequality of (1.1.18) reduces to the inequality of (1.1.12). Thus Theorem 1.4 and Corollary 1.5 generalize the results of Theorems 1.2 and 1.3, even though the process of interpolation is *different* in both cases.

To summarize, this section introduces L-splines and gives representative error bounds for L-spline interpolation. For the extension of these error bounds, we shall find it useful to describe in the next section the idea of interpolating spaces.

1.2. Interpolation spaces and applications. The results of Theorem 1.4 and Corollary 1.5 can be extended to more general spaces, using the idea of *interpolation spaces* (cf. Butzer and Berens [1.16, Chap. 3]), which we briefly describe.

Let X_0 and X_1 be two Banach spaces with norms $\|\cdot\|_0$ and $\|\cdot\|_1$, respectively, which are contained in a linear Hausdorff space $\mathscr{X}$, such that the identity mapping of X_i in $\mathscr{X}$ is continuous for $i = 0$ and $i = 1$. If $X_0 + X_1 \equiv \{f \in \mathscr{X} : f = f_0 + f_1,$ where $f_i \in X_i, i = 0, 1\}$, then $X_0 \cap X_1$ and $X_0 + X_1$ are Banach spaces under the norms:

$$\|f\|_{X_0 \cap X_1} \equiv \max\{\|f\|_0 ; \|f\|_1\},$$
(1.2.1)
$$\|f\|_{X_0 + X_1} \equiv \inf\{\|f_0\|_0 + \|f_1\|_1 : f = f_0 + f_1, f_i \in X_i, i = 0, 1\}.$$

It follows that

$$(1.2.2) \qquad X_0 \cap X_1 \subset X_i \subset X_0 + X_1 \subset \mathscr{X}, \qquad\qquad i = 0, 1,$$

where inclusion is understood in this section to mean that the identity mapping is continuous. Any Banach space $X \subset \mathscr{X}$ is said to be an *intermediate space* of X_0 and X_1 if it satisfies the inclusion

$$(1.2.3) \qquad X_0 \cap X_1 \subset X \subset X_0 + X_1 \subset \mathscr{X}.$$

We now give Peetre's real-variable method (cf. [1.16] and Peetre [1.17]) for constructing intermediate spaces of X_0 and X_1. For each positive t and each $f \in (X_0 + X_1)$, define

$$(1.2.4) \qquad K(t, f) = \inf\{\|f_0\|_0 + t\|f_1\|_1 : f = f_0 + f_1, f_i \in X_i, i = 0, 1\}.$$

Then, for any θ with $0 < \theta < 1$ and any extended real number q with $1 \leqq q \leqq \infty$, let $(X_0, X_1)_{\theta,q}$ be the subset of all $f \in (X_0 + X_1)$ for which the norm

$$(1.2.5) \qquad \|f\|_{(X_0, X_1)_{\theta,q}} \equiv \begin{cases} \left\{ \displaystyle\int_0^\infty [t^{-\theta} K(t, f)]^q \, \frac{dt}{t} \right\}^{1/q}, & 1 \leqq q < \infty, \\[2ex] \sup_{t>0} t^{-\theta} K(t, f), & q = \infty, \end{cases}$$

is finite. Then, the following is known (cf. [1.16, p. 168] and [1.17]).

THEOREM 1.6. *For $0 < \theta < 1$, $1 \leqq q \leqq \infty$, $(X_0, X_1)_{\theta,q}$ is a Banach space which is an intermediate space of X_0 and X_1, and thus satisfies* (1.2.3). *In particular, $(X, X)_{\theta,q} = X$ for any Banach space X.*

Next, let Y_0 and Y_1 be two Banach spaces continuously contained (with respect to the identity mapping) in the linear Hausdorff space $\mathscr{Y}$, and let T denote any linear transformation from $(X_0 + X_1)$ to $(Y_0 + Y_1)$ for which

$$(1.2.6) \qquad \|Tf\|_i \leqq M_i \|f\|_i \quad \text{for all} \quad f \in X_i, \qquad\qquad i = 0, 1,$$

i.e., T is a bounded linear transformation from X_i to Y_i with norm at most M_i, $i = 0, 1$. Again, the following is known (cf. [1.16, p. 180] and [1.17]).

THEOREM 1.7. *Let T be any linear transformation from $(X_0 + X_1)$ to $(Y_0 + Y_1)$ which satisfies (1.2.6). Then, for any $0 < \theta < 1$, $1 \leq q \leq \infty$, T is a bounded linear transformation from the intermediate space $(X_0, X_1)_{\theta,q}$ to the intermediate space $(Y_0, Y_1)_{\theta,q}$, whose norm*

$$M \equiv \sup_{\|f\|_{(X_0, X_1)_{\theta,q}} = 1} \|Tf\|_{(Y_0, Y_1)_{\theta,q}}$$

satisfies

(1.2.7)
$$M \leq M_0^{1-\theta} M_1^\theta.$$

For our purposes here in extending error bounds for L-spline interpolation in one variable, it is sufficient to define the *Besov spaces* $B_p^{\sigma,q}[a, b]$ as spaces intermediate to Sobolev spaces:

(1.2.8)
$$(L_p[a, b], W_p^n[a, b])_{\theta,q} \equiv B_p^{\theta n, q}[a, b],$$

where $0 < \theta < 1$, and $1 \leq p, q \leq \infty$. Some relationships between Besov spaces and Sobolev spaces are given in the following theorem (cf. Grisvard [1.18] and Peetre [1.19]).

THEOREM 1.8. *If $1 \leq p \leq \infty$ and m is a positive integer, then*

(1.2.9)
$$B_p^{m,1}[a, b] \subset W_p^m[a, b] \subset B_p^{m,\infty}[a, b].$$

If $1 \leq q_1 < q_2 \leq \infty$, $1 \leq p \leq \infty$, and $\sigma > 0$, then

(1.2.10)
$$B_p^{\sigma,q_1}[a, b] \subset B_p^{\sigma,q_2}[a, b].$$

If $0 < \sigma_2 < \sigma_1$, $1 \leq q_1, q_2 \leq \infty$, and $1 \leq p \leq \infty$, then

(1.2.11)
$$B_p^{\sigma_1,q_1}[a, b] \subset B_p^{\sigma_2,q_2}[a, b].$$

If $\sigma_0 \neq \sigma_1$, $0 < \theta < 1$, $1 \leq q_0, q_1 \leq \infty$, and $1 \leq p \leq \infty$, then with equivalence of norms,

(1.2.12) $\quad (B_p^{\sigma_0,q_0}[a, b], B_p^{\sigma_1,q_1}[a, b])_{\theta,q} = B_p^{\sigma,q}[a, b], \quad \sigma = \theta\sigma_1 + (1 - \theta)\sigma_0,$

and for integer values of σ_i, either of the spaces $B_p^{\sigma_i,q_i}[a, b]$ in (1.2.12) can be replaced by $W_p^{\sigma_i}[a, b]$. In particular,

(1.2.13)
$$(W_p^m[a, b], W_p^{2m}[a, b])_{\theta,q} = B_p^{\sigma,q}[a, b], \quad \sigma = (1 + \theta)m.$$

It is important to note here that the theory of interpolation spaces provides a useful tool in *higher* dimensions as well, even though our interest at the moment is specifically one-dimensional. In particular, if Ω is a bounded region in R^n, Theorems 1.6 and 1.7 apply directly to the intermediate Besov spaces $B_p^{\sigma,q}(\Omega)$ between the Sobolev spaces $W_p^m(\Omega)$ and $W_p^{m'}(\Omega)$, where m and m' are nonnegative integers and $1 \leq p \leq \infty$. In addition, for the Hilbert space case of $p = 2$, one can also determine intermediate spaces $W_2^\sigma(\Omega)$ to $W_2^m(\Omega)$ and $W_2^{m'}(\Omega)$ for noninteger σ which are also Hilbert spaces. This will be useful in § 6.1.

We now explicitly show how the theory of interpolation spaces can be used to extend the L-spline error bounds of Theorems 1.2 and 1.3. For j a nonnegative integer with $0 \leq j \leq m - 1$ and $2 \leq q \leq \infty$, define the linear transformation $T : W_2^m[a, b] \to W_q^j[a, b]$ by means of

$$(1.2.14) \qquad\qquad\qquad Tg = g - s,$$

where s is the unique $\mathrm{Sp}(L, \Delta, z)$-interpolant of g in the sense of (1.1.6). With this definition of T and the definition of the Sobolev norm in (1.1.3), the error bounds (1.1.10) and (1.1.12) respectively can be expressed as

$$(1.2.15) \quad \begin{cases} \|Tg\|_{W_q^j[a,b]} \leq K\pi^{m-j-1/2+1/q}\|g\|_{W_2^m[a,b]} & \text{for} \quad g \in W_2^m[a, b], \\[2mm] \|Tg\|_{W_q^j[a,b]} \leq K\pi^{2m-j-1/2+1/q}\|g\|_{W_2^{2m}[a,b]} & \text{for} \quad g \in W_2^{2m}[a, b]. \end{cases}$$

Thus, if $Y_0 = Y_1 \equiv W_q^j[a, b]$, and $X_0 \equiv W_2^m[a, b]$ and $X_1 = W_2^{2m}[a, b]$, then T is from (1.2.15) a bounded linear transformation from X_i to Y_i with norm at most M_i, $i = 0, 1$, where

$$(1.2.16) \qquad M_0 = K\pi^{m-j-1/2+1/q}, \quad M_1 = K\pi^{2m-j-1/2+1/q}.$$

Hence, as $(X_0, X_1)_{\theta,r} = (W_2^m[a, b], W_2^{2m}[a, b])_{\theta,r} = B_2^{\sigma,r}[a, b], \sigma \equiv (1 + \theta)m$, from (1.2.13), and as $(Y_0, Y_1)_{\theta,q} = W_q^j[a, b]$, then from Theorem 1.7, T is a bounded linear transformation from $B_2^{\sigma,r}[a, b]$ to $W_q^j[a, b]$ with norm at most

$$M_0^{1-\theta} \cdot M_1^{\theta} = K\pi^{\sigma-j-1/2+1/q}, \quad \sigma = (1 + \theta)m,$$

i.e.,

$$(1.2.17) \qquad \|Tg\|_{W_q^j[a,b]} \equiv \|g - s\|_{W_q^j[a,b]} \leq K\pi^{\sigma-j-1/2+1/q}\|g\|_{B_2^{\sigma,r}[a,b]}$$

for any $m < \sigma < 2m$ and any $1 \leq r \leq \infty$.

The error bounds of (1.2.17) for L-spline interpolation, while extending the results of Theorems 1.2 and 1.3, were obtained by interpolating the *right*-hand sides of (1.1.10) and (1.1.12). On the other hand, the error bounds of (1.1.10) and (1.1.12) also hold for different values of j, and this permits analogous interpolation of the *left*-hand sides of (1.1.10) and (1.1.12). The combination of these results can be formulated (cf. Hedstrom and Varga [1.20]) as the following theorem.

THEOREM 1.9. *Let $f \in B_2^{\sigma_2,r}[a, b]$, where $m < \sigma_2 < 2m$. Then, if s is the unique element in $\mathrm{Sp}(L, \Delta, z)$ interpolating f in the sense of* (1.1.6),

$$(1.2.18) \qquad \|f - s\|_{B_p^{\sigma,q}[a,b]} \leq K\pi^{\sigma_2-\sigma-1/2+1/p}\|f\|_{B_2^{\sigma_2,r}[a,b]}$$

for any $0 < \sigma < m - 1/2 + 1/p$, where $2 \leq p \leq \infty$, and $1 \leq r, q \leq \infty$.

Since $B_p^{j,1}[a, b] \subset W_p^j[a, b]$ from (1.2.9), i.e., $\|f\|_{W_p^j[a,b]} \leq K\|f\|_{B_p^{j,1}[a,b]}$, and since from (1.1.3), $\|D^j f\|_{L_p[a,b]} \leq \|f\|_{W_p^j[a,b]}$, we have from (1.2.18) for the choice $\sigma = j$ and $q = 1$ the immediate consequence of the following corollary.

COROLLARY 1.10. *With the hypotheses of Theorem 1.9,*

$$(1.2.19) \quad \|D^j(f - s)\|_{L_p[a,b]} \leq \|f - s\|_{W_p^j[a,b]} \leq K\pi^{\sigma_2-j-1/2+1/p}\|f\|_{B_2^{\sigma_2,r}[a,b]}$$

for $0 \leqq j \leqq m - 1$ *and any* $2 \leqq p \leqq \infty$. *In particular, if* $f \in W_2^\sigma[a, b]$ *with* $m \leqq \sigma$ $\leqq 2m$, *then for* $0 \leqq j \leqq m$,

$$(1.2.20) \qquad \|D^j(f - s)\|_{L_2[a,b]} \leqq \|f - s\|_{W_2^j[a,b]} \leqq K\pi^{\sigma-j}\|f\|_{W_2^\sigma[a,b]}.$$

The extension in Theorem 1.9 of Theorems 1.2 and 1.3 can be further generalized if we apply the theory of interpolation spaces to Corollary 1.5, where Lagrange interpolation polynomials are used to define interpolation (cf. (1.1.15)). From (1.1.18) we obtain the following result.

THEOREM 1.11. *Given any* $f \in B_r^{\sigma,q}[a, b]$, $1 < \sigma < 2m$, $1 \leqq q, r \leqq \infty$, *and given* $\Delta \in \mathscr{P}_\sigma(a, b)$ *with at least* $2m$ *knots, let* s *be the unique element in* $\mathrm{Sp}(L, \Delta, z)$ *which interpolates* f *in the sense of* (1.1.15). *Then, for* $\max(r, 2) \leqq p \leqq \infty$,

$$(1.2.21) \qquad \|f - s\|_{B_p^{\tau,q'}[a,b]} \leqq K\pi^{\sigma-\tau+1/p+\min(-1/r,-1/2)}\|f\|_{B_r^{\sigma,q}[a,b]}$$

for $0 < \tau < \min(\sigma - 1, 2m - z)$, $1 \leqq q', p \leqq \infty$.

To summarize, this section briefly discusses the theory of interpolation spaces, and gives applications of interpolation space theory to extensions of known error bounds for L-spline interpolation. This theory is a very useful tool in numerical analysis, and will be mentioned again in § 6.1.

REFERENCES

[1.1] I. J. SCHOENBERG, *Contributions to the problem of approximation of equidistant data by analytic functions, Parts A and B*, Quart. Appl. Math., 4 (1946), pp. 45–99, 112–141.

[1.2] J. H. AHLBERG, E. N. NILSON AND J. L. WALSH, *The Theory of Splines and Their Applications*, Academic Press, New York, 1967.

[1.3] T. N. E. GREVILLE, editor, *Theory and Applications of Spline Functions*, Academic Press, New York, 1969.

[1.4] I. J. SCHOENBERG, editor, *Approximations with Special Emphasis on Spline Functions*, Academic Press, New York, 1969.

[1.5] T. N. E. GREVILLE, *Interpolation by generalized spline functions*, MRC Tech. Summ. Rep. 476, Mathematics Research Center, United States Army, University of Wisconsin, Madison, 1964.

[1.6] M. H. SCHULTZ AND R. S. VARGA, *L-splines*, Numer. Math., 10 (1967), pp. 345–369.

[1.7] G. BIRKHOFF, M. H. SCHULTZ AND R. S. VARGA, *Piecewise Hermite interpolation in one and two variables with applications to partial differential equations*, Ibid., 11 (1968), pp. 232–256.

[1.8] MICHAEL GOLOMB, *Approximation by periodic spline interpolation on uniform meshes*, J. Approx. Theory, 1 (1968), pp. 26–65.

[1.9] I. BABUŠKA, M. PRÁGER AND E. VITÁSEK, *Numerical Processes in Differential Equations*, Interscience, London, 1966.

[1.10] G. G. LORENTZ, *Approximation of Functions*, Holt, Rinehart and Winston, New York, 1966.

[1.11] J. P. AUBIN, *Interpolation et approximation optimales; Spline functions*, J. Math. Anal. Appl., 24 (1968), pp. 1–24.

[1.12] MICHAEL GOLOMB, *Splines, n-widths and optimal approximations*, MRC Tech. Summ. Rep. 784, Mathematics Research Center, United States Army, University of Wisconsin, Madison, 1967.

[1.13] BLAIR SWARTZ AND RICHARD S. VARGA, *Error bounds for spline and L-spline interpolation*, J. Approx. Theory, to appear.

[1.14] PHILIP J. DAVIS, *Interpolation and Approximation*, Blaisdell, New York, 1963.

[1.15] M. H. SCHULTZ, *L^∞-multivariate approximation theory*, SIAM J. Numer. Anal., 6 (1969), pp. 161–183.

[1.16] P. L. BUTZER AND H. BERENS, *Semi-Groups of Operators and Approximations*, Springer-Verlag, New York, 1967.

[1.17] J. PEETRE, *Introduction to Interpolation*, Lecture notes, Department of Mathematics, Lund, 1966. (In Swedish.)

[1.18] P. GRISVARD, *Commutativité de deux foncteurs d'interpolation et applications*, J. Math. Pures Appl., 45 (1966), pp. 143–290.

[1.19] J. PEETRE, *Espaces d'interpolation, généralisations, applications*, Rend. Sem. Mat. Fiz. Milano, 34 (1964), pp. 133–164.

[1.20] GERALD W. HEDSTROM AND RICHARD S. VARGA, *Application of Besov spaces to spline approximation*, J. Approx. Theory, to appear.

CHAPTER 2

Generalizations of *L*-Splines

2.1. *Lg*-splines. There are a variety of generalizations of *L*-splines and it is of interest to see how they extend the *L*-spline theory. We begin this section with results of Jerome and Schumaker [2.1].

Let $\Lambda = \{\lambda_i\}_{i=1}^k$ be any set of linearly independent, bounded linear functionals on $W_2^m[a, b]$, and let $\mathbf{r} = (r_1, r_2, \cdots, r_k)^T$ denote any vector of real Euclidean k-space, R^k. If L is the linear differential operator of (1.1.4), then (cf. [2.1]) $s \in W_2^m[a, b]$ is an *Lg-spline* interpolating $\mathbf{r}$ with respect to Λ, i.e., $\lambda_i(s) = r_i$, $1 \leq i \leq k$, provided it solves the following minimization problem:

$$\|Ls\|_{L_2[a,b]} = \inf\{\|Lf\|_{L_2[a,b]} : f \in U_\Lambda(\mathbf{r})\},$$

(2.1.1)
$$\text{where} \quad U_\Lambda(\mathbf{r}) \equiv \{f \in W_2^m[a, b] : \lambda_i(f) = r_i, 1 \leq i \leq k\}.$$

The relationship with *L*-splines in (1.1.9) of Theorem 1.1 is clear; while the linear differential operator L remains unchanged, the manner of interpolation, now by means of Λ, is generalized. For notation, the space of all *Lg*-splines such that s satisfies (2.1.1) for some $\mathbf{r} \in R^k$ is denoted by $\mathrm{Sp}(L, \Lambda)$.

Based on results of Golomb [2.2], Jerome and Schumaker [2.1] have proved, in the spirit of Anselone and Laurent [2.3], the following theorem.

THEOREM 2.1. *Given any* $\mathbf{r} \in R^k$, *there exists an* $s \in W_2^m[a, b]$ *satisfying* (2.1.1). *A function* $s \in U_\Lambda(\mathbf{r})$ *satisfies* (2.1.1) *if and only if*

(2.1.2)
$$\int_a^b Ls \cdot Lg \, dx = 0 \quad \text{for all} \quad g \in U_\Lambda(\mathbf{0}).$$

Moreover, (2.1.1) *possesses a unique solution if and only if* $\mathfrak{N} \cap U_\Lambda(\mathbf{0}) = \{0\}$, *where* $\mathfrak{N}$ *is the null space of* L. *Finally,* $\mathrm{Sp}(L, \Lambda)$ *is a linear space of dimension* $k + \dim\{\mathfrak{N} \cap U_\Lambda(\mathbf{0})\}$ *in* $W_2^m[a, b]$.

We now assume that each $\lambda_i \in \Lambda$ is of the form $\lambda_i(f) = D^{j_i}f(x_i)$, where $0 \leq j_i \leq m - 1$ and $x_i \in [a, b]$. Such a $\Lambda = \{\lambda_i\}_{i=1}^k$ generates a *Hermite–Birkhoff problem*. For such Hermite–Birkhoff problems, the solution s of the minimization problem (2.1.1) satisfies, as in (1.1.5),

(2.1.3)
$$L^*Ls(x) = 0, \quad x \in (a, b) - \{x_i\}_{i=1}^k.$$

Next, one can assign a nonnegative integer $\tau(\Lambda)$, which counts the number of consecutive derivative point functionals in Λ (for details, see Jerome and Varga [2.4]). With this, the following can readily be shown (cf. [2.4]).

THEOREM 2.2. *If* $\Lambda = \{\lambda_i\}_{i=1}^k$ *generates a Hermite–Birkhoff problem with partition* $\Delta \in \mathscr{P}_\sigma(a, b)$, *assume that* $\tau(\Lambda) \geqq m$, *assume that* $\mathfrak{N} \cap U_\Lambda(0) = \{0\}$, *and assume that the second integral relation holds for* Λ, *i.e.* (cf. (1.1.11)),

$$(2.1.4) \qquad \int_a^b \{L(g - s)\}^2 \, dx = \int_a^b (g - s)L^*Lg \, dx$$

is valid for any $g \in W_2^{2m}[a, b]$ *and s is the unique Lg-spline which interpolates g in the sense that*

$$(2.1.5) \qquad\qquad \lambda_i(s) = \lambda_i(g), \qquad\qquad 1 \leqq i \leqq k.$$

Then the error bounds of (1.1.10) *of Theorem* 1.2, *as well as those of* (1.1.12) *of Theorem* 1.3 *are valid.*

More broadly interpreted, the result of Theorem 2.2 can be clearly extended to Besov spaces, exactly as in Theorem 1.9 and Corollary 1.10, with *identical* error bounds, thereby generalizing the results of [2.4]. Thus, Lg-splines offer generalizations in the area of interpolation (cf. (2.1.5)), but do not generalize the type of differential operator, L, considered.

2.2 γ-splines.

The next generalization considered here is due to Schultz [2.5] and Lucas [2.6]. If

$$(2.2.1) \qquad Eu(x) \equiv \sum_{j=0}^m (-1)^j D^j \{p_j(x)D^j u(x)\},$$

where $p_j \in W_2^j[a, b] \cap L_\infty[a, b], 0 \leqq j \leqq m$, and $p_m(x) \geqq \delta > 0$ in $[a, b]$, assume that E is $\mathring{W}_2^m[a, b]$-*elliptic*, i.e., there exists a constant $\gamma > 0$ such that

$$(2.2.2) \qquad \|u\|_{W_2^m[a,b]}^2 \leqq \gamma^2 \int_a^b \left\{ \sum_{j=0}^m p_j(D^j u)^2 \right\} dx \equiv \gamma^2 e(u, u) \quad \text{for all} \quad u \in \mathring{W}_2^m[a, b],$$

where $\mathring{W}_2^m[a, b]$ denotes the subspace of functions $u(x)$ of $W_2^m[a, b]$ of § 1.1 satisfying the homogeneous boundary conditions $D^k u(a) = D^k u(b) = 0$, $0 \leqq k \leqq m - 1$. As in § 1.1, let $\Delta \in \mathscr{P}(a, b)$, and let z again be a positive integer satisfying $1 \leqq z \leqq m$. Then, $S(E, \Delta, z)$, *the γ-spline space*, is the collection of real-valued functions w defined on $[a, b]$ such that, relative to Δ (cf. (1.1.5)),

$$Ew(x) = 0 \quad \text{almost everywhere in each subinterval} \quad (x_i, x_{i+1}),$$

$$(2.2.3) \qquad\qquad\qquad\qquad\qquad\qquad\qquad\qquad\qquad\qquad 0 \leqq i \leqq N - 1,$$

$$D^k w(x_i -) = D^k w(x_i +) \quad \text{for} \quad 0 \leqq k \leqq 2m - 1 - z, \qquad 0 < i < N.$$

Given any $g \in C^{m-1}[a, b]$, it is easily seen (cf. [2.5]) that there exists a unique $s \in S(E, \Delta, z)$ which interpolates g in the sense of (1.1.6), i.e.,

$$
\begin{aligned}
D^j(g - s)(x_i) &= 0, &\qquad 0 \leqq j \leqq z - 1, \quad 0 < i < N, \\
D^j(g - s)(a) &= D^j(g - s)(b) = 0, &\qquad 0 \leqq j \leqq m - 1.
\end{aligned}
$$

$$(2.2.4)$$

Because the interpolation of (2.2.4) at the boundaries ensures the second integral relation (cf. (1.1.11)), the following contains the upper bounds of Theorems 2.4–2.7 of Schultz [2.5].

THEOREM 2.3. *Given any* $g \in C^{m-1}[a, b]$ *and any* $\Delta \in \mathscr{P}(a, b)$, *let s be the unique element in* $S(E, \Delta, z)$ *which interpolates g in the sense of* (2.2.4). *Then, the error bounds of* (1.1.10) *of Theorem 1.2 are valid. Similarly, if* $g \in W_2^{2m}[a, b]$ *and* $\Delta \in \mathscr{P}_\sigma(a, b)$, *then the bounds of* (1.1.12) *of Theorem 1.3 are valid.*

As in § 2.1, we can more broadly interpret the result of Theorem 2.3, since its extension to Besov spaces, exactly as in Theorem 1.9 and Corollary 1.10, is now immediate, thereby generalizing the results of [2.5] and [2.6].

Noting that the generalization of *Lg*-splines works through more general collections of bounded linear functionals $\Lambda = \{\lambda_i\}_{i=1}^k$, while the generalization of γ-splines works through more general differential operators, one can combine these two ideas simultaneously, and obtain the error bounds of (1.1.10) of Theorem 1.2 and (1.1.12) of Theorem 1.3. This has in fact been considered by Lucas [2.7]. The extension of these results to Besov spaces is also immediate.

2.3. Singular splines. One of the more interesting developments with respect to one-dimensional spline theory is due, in its generalized form, to Jerome and Pierce [2.8]. We first give a brief discussion of the background for this problem. Jamet [2.9], using finite-differences, considered the numerical approximation of the solution of the *singular* boundary value problem:

$$D^2 u(x) + \frac{\sigma}{x} Du(x) = f(x), \qquad\qquad 0 < x < 1,$$

(2.3.1)
$$u(0) = \alpha, \quad u(1) = \beta,$$

where $0 \leqq \sigma < 1$. After a change of variables, this problem can be put into the self-adjoint form:

$$D\{x^\sigma Du(x)\} = g(x), \qquad\qquad 0 < x < 1,$$

(2.3.2)
$$u(0) = u(1) = 0.$$

Using a uniform partition ($\pi = h$) of $[0, 1]$, and a standard three-point difference approximation to (2.3.1), Jamet [2.9] obtained a discrete approximation to the solution of (2.3.1), with an upper bound for the error in a discrete L_∞-norm of the form $Kh^{1-\sigma}$. In Ciarlet, Natterer and Varga [2.10], a variational approximation to the solution of (2.3.2) was found, with a sharp upper bound for the error in the uniform norm of the form $Kh^{2-\sigma}$. For this problem (2.3.2), the variational approximation was made up locally of solutions of

$$D\{x^\sigma Dw(x)\} = 0, \qquad\qquad 0 \leqq \sigma < 1,$$

on each subinterval of the uniform partition Δ_u of $[0, 1]$ with $\pi = h$. In [2.10], the problem of (2.3.2) was generalized to

$$D\{p(x)Du(x)\} = f(x, u(x)), \qquad\qquad 0 < x < 1,$$

(2.3.3)
$$u(0) = u(1) = 0,$$

where it was assumed that

$$p(x) > 0 \quad \text{in} \quad (0, 1),$$

(2.3.4) $$p \in C^1[0, 1], \quad \text{and}$$

$$\frac{1}{p} \in L_1[0, 1],$$

and approximate solutions of (2.3.3) were made up of solutions of

$$D\{p(x)Dw(x)\} = 0$$

on subintervals defined by a partition Δ of $[0, 1]$. While it is true that the above equation can be expressed as $-L^*Lw(x) = 0$, where

$$Lw(x) \equiv \sqrt{p(x)}Dw(x),$$

note that since $p(0)$ can be zero in (2.3.4), these "splines" are not in general L-splines or γ-splines. Generalizations to higher order singular Hermite splines were also considered by Dailey [2.11].

To describe the singular Λ-splines of Jerome and Pierce [2.8] (which generalizes [2.10]), let Λ be the formally self-adjoint operator defined by

(2.3.5) $$\Lambda u(x) \equiv \sum_{j=0}^{m} (-1)^j D^j \{a_j(x)D^j u(x)\},$$

where it is assumed that (cf. (2.3.4))

$$a_m(x) > 0 \quad \text{for all} \quad x \in (a, b),$$

(2.3.6) $$a_j \in C^j[a, b], \qquad\qquad\qquad\qquad 0 \leq j \leq m,$$

$$\frac{1}{a_m} \in L_1[a, b].$$

Next, let H denote the weighted Sobolov space of all real-valued functions f defined on $[a, b]$ such that $D^{m-1}f$ is absolutely continuous and $\sqrt{a_m}D^m f \in L_2[a, b]$, with norm

(2.3.7) $$\|u\|_H^2 \equiv \sum_{j=0}^{m-1} (D^j u(a))^2 + \|\sqrt{a_m}D^m u\|_{L_2[a,b]},$$

and consider the bilinear form $B(u, v)$ associated with Λ:

(2.3.8) $$B(u, v) \equiv \int_a^b \left\{ \sum_{j=0}^{m} a_j D^j u D^j v \right\} dx, \qquad\qquad u, v \in H,$$

on $H \times H$. We assume that the operator Λ is $\mathring{H}$-*elliptic*, i.e., a positive constant ρ exists such that

(2.3.9) $$\|u\|_H^2 \leq \rho B(u, u) \quad \text{for all} \quad u \in \mathring{H},$$

where $\mathring{H}$ denotes the closed subspace of H of all functions f with $D^j f(a) = D^j f(b) = 0$ for $0 \leq j \leq m - 1$. From (2.3.8) and (2.3.9), it readily follows that $\mathring{H}$ is a Hilbert space under the inner product

$$(2.3.10) \qquad (u, v)_D = B(u, v).$$

Next, let $M = \{\lambda_j\}_{j=1}^k$ be any set of bounded linear functionals which are linearly independent on $\mathring{H}$. Then (cf. [2.8]), $s \in \mathring{H}$ is a Λ-*spline* interpolating $\mathbf{r} = (r_1, r_2, \cdots, r_k) \in R^k$ if s solves the minimization problem:

$$(2.3.11) \qquad B(s, s) = \min_{f \in U(r)} B(f, f),$$
$$\text{where} \quad U(\mathbf{r}) = \{ f \in \mathring{H} : \lambda_j(f) = r_j, 1 \leq j \leq k \}.$$

Because of the vanishing boundary data for $f \in \mathring{H}$, it follows that, for any $f \in \mathring{H}$, there exists a *unique* Λ-spline s which interpolates f in the sense that

$$(2.3.12) \qquad \lambda_j(s) = \lambda_j(f), \qquad\qquad 1 \leq j \leq k.$$

As before, $\mathrm{Sp}(\Lambda, M)$, the class of all s which satisfies (2.3.11) for some $\mathbf{r} \in R^k$, is a *linear* space.

In order to obtain error estimates for the interpolation of (2.3.12), we next assume, as in § 2.1, that $M = \{\lambda_j\}_{j=1}^k$ generates a Hermite–Birkhoff problem, i.e., each $\lambda_i \in M$ is of the form $\lambda_i(f) = D^{j_i}f(x_i)$ with $0 \leq j_i \leq m - 1$ and $x_i \in [a, b]$. In this case, any Λ-spline s is a solution of $\Lambda s = 0$ on subintervals defined from M, just as in (2.2.3). Because we are considering $\mathring{H}$, a second-integral relation holds, and the following error bounds can be proved (cf. [2.8]).

THEOREM 2.4. *If M generates a Hermite–Birkhoff problem with partition $\Delta \in \mathscr{P}_\sigma(a, b)$, assume that $\tau(M) \geq n$, and for any $f \in \mathring{H}$, let $s \in \mathring{H}$ be its unique Λ-spline interpolation* (cf. (2.3.12)). *Then, for any $2 \leq q \leq \infty$,*

$$(2.3.13) \qquad \| D^j(f - s) \|_{L_q[a,b]} \leq K \left(\omega_1 \left(\frac{1}{a_m}, \pi \right) \right)^{1/2} \pi^{m - j - 1 + 1/q} \| f \|_D,$$

$$0 \leq j \leq m - 1,$$

where $\| f \|_D^2 = (f, f)_D$ (cf. (2.3.10)), *and where*

$$(2.3.14) \qquad \omega_1 \left(\frac{1}{a_m}, \delta \right) \equiv \sup \left\{ \int_x^{x+t} \frac{d\xi}{a_m(\xi)} : x, x + t \text{ are in } [a, b], |t| \leq \delta \right\}.$$

In addition, if $\Lambda f \in L_2[a, b]$, then for $2 \leq q \leq \infty$,

$$(2.3.15) \qquad \| D^j(f - s) \|_{L_q[a,b]} \leq K \omega_1 \left(\frac{1}{a_m}, \pi \right) \cdot \pi^{2m - j - 3/2 + 1/q} \| \Lambda f \|_{L_2[a,b]}$$

for $0 \leq j \leq m - 1$.

It is worth remarking that the Sobolev space $\mathring{W}_2^j[a, b]$, for $j \geq m$, are all subspaces of $\mathring{H}$. However, to extend the results of Theorem 2.4 via the theory of intermediate spaces, as in the previous theorems of this section, we would necessarily work with spaces intermediate to $\overset{\circ}{H}$ and, say, $W_2^{2m}[a, b]$, and these are *not*, as in previous cases, Besov spaces.

It is also worth noting that the results of Jerome and Pierce [2.8] go beyond the assumption of (2.3.9), i.e., weaker assumptions are made in [2.8] corresponding to (2.3.9), and existence and uniqueness of interpolation plus error bounds for the more general extended Hermite–Birkhoff problem are treated there. Since the case $a_m(x) \geqq \delta > 0$ on $[a, b]$ is not ruled out in (2.3.6), the results of [2.8] thus simultaneously generalize Lg-splines and γ-splines and give, as a special case, the known results for error bounds for spline interpolation of Theorems 2.2 and 2.3.

REFERENCES

[2.1] J. W. Jerome and L. L. Schumaker, *On Lg-splines*, J. Approx. Theory, 2 (1969), pp. 29–49.

[2.2] M. Golomb, *Splines, n-widths, and optimal approximation*, MRC Tech. Summ. Rep. 784, Mathematics Research Center, United States Army, University of Wisconsin, Madison, 1967.

[2.3] P. M. Anselone and P. J. Laurent, *A general method for the construction of interpolating or smoothing spline-functions*, Numer. Math., 12 (1968), pp. 66–82.

[2.4] J. W. Jerome and R. S. Varga, *Generalizations of spline functions and applications to nonlinear boundary value and eigenvalue problems*, Theory and Applications of Spline Functions, T. N. E. Greville, ed., Academic Press, New York, 1969, pp. 103–155.

[2.5] M. H. Schultz, *Elliptic spline functions and the Rayleigh–Ritz–Galerkin method*, Math. Comp., 24 (1970), pp. 65–80.

[2.6] T. R. Lucas, *A generalization of L-splines*, Numer. Math., 15 (1970), pp. 359–370.

[2.7] ———, *A theory of generalized splines with applications to nonlinear boundary value problems*, Thesis, Georgia Institute of Technology, 1970.

[2.8] J. Jerome and J. Pierce, *On spline functions determined by singular self-adjoint differential operators*, J. Approx. Theory, to appear.

[2.9] P. Jamet, *On the convergence of finite-difference approximations to one-dimensional singular boundary-value problems*, Numer. Math., 14 (1970), pp. 355–378.

[2.10] P. G. Ciarlet, F. Natterer and R. S. Varga, *Numerical methods of high-order accuracy for singular nonlinear boundary value problems*, Ibid., 15 (1970), pp. 87–99.

[2.11] J. W. Dailey, *Approximation by spline-type functions and related problems*, Thesis, Case Western Reserve University, 1969.

CHAPTER 3

Interpolation and Approximation Results for Piecewise-Polynomials in Higher Dimensions

3.1. Tensor products of one-dimensional polynomial splines. For many applications, it is desirable to generalize the results of Chapters 1 and 2 for one-dimensional piecewise-polynomial functions or splines to n-dimensional analogues. The easiest of such extensions is obtained by simply considering the *tensor product* of one-dimensional spline spaces. This was considered in Birkhoff, Schultz and Varga [3.1], specifically for the tensor product of Hermite polynomial splines in two-space variables. To describe briefly the results of [3.1], let $\Delta \equiv \Delta_1 \times \Delta_2$, given by

$$\Delta_1 : a = x_0 < x_1 < \cdots < x_N = b,$$

(3.1.1)

$$\Delta_2 : c = y_0 < y_1 < \cdots < y_{N'} = d,$$

denote a partition of the rectangle $\bar{\Omega} \equiv [a, b] \times [c, d]$ in R^2. If $H^{(m)}(\Delta; \Omega)$ is the set of all real-valued piecewise-polynomial functions $w(x, y)$ defined on Ω such that $D^{(i,j)}w \equiv D_x^i D_y^j w$ is continuous in $\bar{\Omega}$ for all $0 \leq i, j \leq m - 1$, and such that $w(x, y)$ is a polynomial of degree $2m - 1$ in *each* of the variables x and y in each sub-rectangle $[x_i, x_{i+1}] \times [y_j, y_{j+1}]$ defined on Ω by Δ, we can define a unique interpolation s in $H^{(m)}(\Delta; \Omega)$ of a real-valued function f such that $D^{(p,q)}f$ is continuous in $\bar{\Omega}$ for all $0 \leq p, q \leq m - 1$, by means of

$$(3.1.2) \qquad D^{(i,j)}s(x_k, y_l) = D^{(i,j)}f(x_k, y_l),$$

$$0 \leq i, j \leq m - 1, \quad 0 \leq k \leq N, \quad 0 \leq l \leq N'.$$

If π_1, $\underline{\pi}_1$, π_2 and $\underline{\pi}_2$ have the analogous meaning (cf. §1.1) for the partition $\Delta = \Delta_1 \times \Delta_2$ of $\bar{\Omega}$, we assume that Δ is an element of $\mathscr{P}_\sigma(\Omega)$, with finite $\sigma \geq 1$, i.e., (cf. §1.1),

$$(3.1.3) \qquad \pi \equiv \max_{i=0,1} \pi_i, \quad \underline{\pi} \equiv \min_{i=0,1} \underline{\pi}_i, \qquad \frac{\pi}{\underline{\pi}} \leq \sigma.$$

Then, using the idea of the Peano kernel theorem (cf. Sard [3.2]) in two-space variables, the following was shown in [3.1]. We use the notation $S_2^p(\Omega)$ to denote the set of all real-valued functions f defined on Ω such that $D^{(p-i,i)}f \in L_2(\Omega)$ for all $0 \leq i \leq p$, and such that $D^{(i,j)}f$ is continuous in $\bar{\Omega}$ for all $0 \leq i + j < p$.

THEOREM 3.1. *Given* $f \in S_2^{2m}(\Omega)$, *where* $\Omega = (a, b) \times (c, d)$, *and given* Δ

$= \Delta_1 \times \Delta_2 \in \mathscr{P}_\sigma(\Omega)$, *let* $s \in H^{(m)}(\Delta; \Omega)$ *be the unique interpolation of* f *in the sense of* (3.1.2). *Then*

$$(3.1.4) \qquad \|D^{(h,l)}(f - s)\|_{L_2(\Omega)} \leqq K\pi^{2m-h-l} \sum_{j=0}^{2m} \|D^{(j,2m-j)}f\|_{L_2(\Omega)}$$

for all $0 \leqq h, l \leqq m$ *with* $0 \leqq h + l \leqq m$.

Because the Hermite interpolation of (3.1.2) is *local*, the result of Theorem 3.1 is actually valid for any *rectangular polygon*, i.e., any polygon whose sides are parallel to the coordinate axes in the plane, such as an L-shaped region.

For our future needs, we now introduce the following notation. With n any positive integer, let Ω be a bounded region in Euclidean n-space, R^n. We assume that the bounded region Ω in R^n satisfies a *restricted cone condition* (cf. Agmon [3.3, p. 11]), i.e., there exist a finite open cover $\{\mathcal{O}_i\}_{i=1}^m$ of the boundary $\partial\Omega$ of Ω, where each $\mathcal{O}_i$ is an open subset of R^n, and associated open truncated cones $\{C_i\}_{i=1}^m$ with vertices at the origin such that for any i and any $x \in \mathcal{O}_i \cap \Omega$, then $x + C_i \equiv \{w : w = x + y, \text{ where } y \in C_i\}$ lies in Ω. Next, if $\alpha = (\alpha_1, \alpha_2, \cdots, \alpha_n)$ is any n-tuple of nonnegative integers, then

$$D^\alpha \equiv \frac{\partial^{\alpha_1 + \alpha_2 + \cdots + \alpha_n}}{\partial x_1^{\alpha_1} \partial x_2^{\alpha_2} \cdots \partial x_n^{\alpha_n}}$$

denotes the differential operator of order $|\alpha| \equiv \sum_{i=1}^n \alpha_i$. The space of all real-valued functions which have continuous derivatives of all orders α with $|\alpha| \leqq m$ in Ω is denoted by $C^m(\Omega)$. The space $C_0^\infty(\Omega)$ is the collection of all infinitely differentiable functions u in Ω which vanish identically outside some compact set contained in Ω. Similarly, $C_0^\infty(R^n)$ is the collection of all infinitely differentiable functions in R^n which vanish identically outside some compact set in R^n.

The Sobolev spaces $W_2^m(\Omega)$ and $W_\infty^m(\Omega)$, m a nonnegative integer, are then defined as the respective completions of $C^\infty(\overline{\Omega})$ in the norm:

$$\|u\|_{W_2^m(\Omega)}^2 \equiv \sum_{|\alpha| \leqq m} \|D^\alpha u\|_{L_2(\Omega)}^2,$$

$$(3.1.5)$$

$$\|u\|_{W_\infty^m(\Omega)} \equiv \sum_{|\alpha| \leqq m} \|D^\alpha u\|_{L_\infty(\Omega)}.$$

Similarly, $W_2^m(R^n)$ and $W_\infty^m(R^n)$ are the completions of $C_0^\infty(R^n)$ in the above norms, and $\mathring{W}_2^m(\Omega)$ is the completion of $C_0^\infty(\Omega)$ in the first norm of (3.1.5).

The result of Theorem 3.1 can be interpreted in terms of Sobolev norms as follows.

COROLLARY 3.2. *With the hypotheses of Theorem* 3.1,

$$(3.1.6) \qquad \|f - s\|_{W_2^k(\Omega)} \leqq K\pi^{2m-k} \sum_{j=0}^{2m} \|D^{(j,2m-j)}f\|_{L_2(\Omega)}$$

for any k *with* $0 \leqq k \leqq m$.

There are two inherent shortcomings of Theorem 3.1 and its Corollary 3.2. One is that the results as stated pertain only to rectangular polygons in R^2, and not to higher dimensions. The other is that the function space $S_2^{2m}(\Omega)$ used in these results is not what one would expect, namely the usual Sobolev space $W_2^{2m}(\Omega)$. These shortcomings of [3.1] have been more than adequately covered by Bramble and Hilbert's generalization in [3.4], [3.5] and [3.6], which we now describe.

Consider any closed hypercube $\bar{\Omega}$ in R^n, with its 2^n vertices denoted by x_i, $1 \leq i \leq 2^n$. For $u \in C^{2m-1}(\bar{\Omega})$, the *mth Hermite interpolation* u_m of u in $\bar{\Omega}$ is defined as a polynomial of degree $2m - 1$ in each of its n variables which satisfies at each vertex x_i

$$(3.1.7) \qquad D^\gamma u_m(x_i) = \begin{cases} D^\gamma u(x_i), & |\gamma| < 2m, \\ 0, & |\gamma| \geq 2m, \end{cases}$$

for any $\gamma = (\gamma_1, \cdots, \gamma_n)$ with $0 \leq \gamma_j \leq m - 1$ for all $1 \leq j \leq n$. Because (3.1.7) is a nonsingular linear system of $(2m)^n$ equations in $(2m)^n$ unknowns, it is readily seen that the function u_m, so defined, is unique. This approach, in fact, generalizes the Hermite-interpolation of (3.1.2) in two dimensions.

Next, let R^n be decomposed into hypercubes *with sides of length h*, i.e., $R^n = \bigcup_i \bar{\Omega}_i$, where $\bar{\Omega}_i \cap \bar{\Omega}_j$ is, for any $i \neq j$, either empty or a part of the boundary of $\bar{\Omega}_i$. Because the interpolation of (3.1.7) is *local*, then given any $u \in C^{2m-1}(R^n)$, a unique interpolant u_m of u can be found which satisfies (3.1.7) at every vertex of every Ω_i. As is readily seen from (3.1.7), $D^\alpha u_m$ is continuous in R^n for any $\alpha = (\alpha_1, \cdots, \alpha_n)$ with $0 \leq \alpha_i \leq m - 1$, $i = 1, 2, \cdots, n$.

Unlike the one-dimensional case, an arbitrary function $u \in W_2^{2m}(R^n)$ need *not* have well-defined derivatives at the vertices of the Ω_i for the interpolation procedure of (3.1.7). Thus, it is necessary to *smooth* or *mollify* u to obtain a $u_h \in C_0^\infty(R^n)$ for which the interpolation of (3.1.7) is meaningful. (This is the analogue of the use of Lagrange polynomial interpolation in § 1.1.) It has been shown by Bramble and Hilbert that a mollified $u_h \in C_0^\infty(R^n)$, for $u \in W_2^{2m}(R^n)$, can be found such that for all $h > 0$, there exists a constant C, independent of h and u, with

$$(3.1.8) \qquad \|D^\alpha(u - u_h)\|_{L_2(R^n)} \leq Ch^{2m-|\alpha|}\|u\|_{W_2^{2m}(R^n)}, \qquad |\alpha| \leq 2m.$$

Next, with $u_h \in C_0^\infty(R^n)$, let $u_{h,m}$ be the Hermite interpolation of u_h over hypercubes of side h, in the sense of (3.1.7). Then, based on a generalization of the Peano kernel theorem, Bramble and Hilbert have shown that

$$(3.1.9) \qquad \|D^\alpha(u_h - u_{h,m})\|_{L_2(R^n)} \leq Ch^{2m-|\alpha|}\|u_h\|_{W_2^{2m}(R^n)}$$

for any $\alpha = (\alpha_1, \alpha_2, \cdots, \alpha_n)$ with $|\alpha| \leq 2m$ and $0 \leq \alpha_i \leq m - 1$ for all $1 \leq i \leq n$. Thus, combining (3.1.8) and (3.1.9) gives

$$(3.1.10) \qquad \|D^\alpha(u - u_{h,m})\|_{L_2(R^n)} \leq Ch^{2m-|\alpha|}\|u\|_{W_2^{2m}(R^n)}$$

for any α with $|\alpha| \leq 2m$ and $0 \leq \alpha_i \leq m - 1$ for all $1 \leq i \leq n$.

To extend this result of (3.1.10) to a *bounded* region $\Omega \subset R^n$, we use the *Calderón extension theorem* (cf. [3.3, p. 171]). Specifically, if Ω satisfies a restricted cone property, there exists a *bounded* linear transformation $\mathscr{E}: W_2^{2m}(\Omega) \to W_2^{2m}(R^n)$ with $\mathscr{E}v = v$ on Ω, i.e., for some positive constant C,

$$(3.1.11) \qquad \|\mathscr{E}v\|_{W_2^{2m}(R^n)} \leqq C\|v\|_{W_2^{2m}(\Omega)} \quad \text{for all} \quad v \in W_2^{2m}(\Omega).$$

Thus, given any $u \in W_2^{2m}(\Omega)$, then $\mathscr{E}u$ is an element of $W_2^{2m}(R^n)$, and (3.1.10) then can be applied to $\mathscr{E}u$, i.e., with (3.1.10) and (3.1.11),

$$\|D^\alpha(\mathscr{E}u - (\mathscr{E}u)_{h,m})\|_{L_2(R^n)} \leqq Ch^{2m-|\alpha|}\|\mathscr{E}u\|_{W_2^{2m}(R^n)}$$

$$\leqq C'h^{2m-|\alpha|}\|u\|_{W_2^{2m}(\Omega)}.$$

However, since by definition $\|v\|_{L_2(R^n)} \geqq \|v\|_{L_2(\Omega)}$ for any $v \in L_2(R^n)$ and since $\mathscr{E}u = u$ on Ω, it necessarily follows that the above inequality gives us that, for any α with $|\alpha| \leqq 2m$ and $0 \leqq \alpha_i \leqq m - 1$ for all $1 \leqq i \leqq n$,

$$(3.1.12) \qquad \inf_{w \in H_h^{(m)}(\Omega)} \|D^\alpha(u - w)\|_{L_2(\Omega)} \leqq Ch^{2m-|\alpha|}\|u\|_{W_2^{2m}(\Omega)},$$

where if $H_h^{(m)}(R^n)$ is the subspace of $C^{m-1}(R^n)$ of all functions which are polynomials of degree $2m - 1$ in each variable on each hypercube Ω_i of $R^n = \bigcup_i \bar{\Omega}_i$ of side h, then $H_h^{(m)}(\Omega)$ is simply the *restriction* of $H_h^{(m)}(R^n)$ to Ω. It turns out that not all terms in

$$\left\{ \sum_{|\alpha| \leqq 2m} \|D^\alpha u\|_{L_2(\Omega)}^2 \right\}^{1/2}$$

of $\|u\|_{W_2^{2m}(\Omega)}$ are needed, as conjectured by G. Birkhoff. The improved form of (3.1.12) of Bramble and Hilbert [3.4] is given in the following theorem.

THEOREM 3.3. *For any* $u \in W_2^{2m}(\Omega)$,

$$(3.1.13) \qquad \inf_{w \in H_h^{(m)}(\Omega)} \|D^\alpha(u - w)\|_{L_2(\Omega)} \leqq Ch^{2m-|\alpha|} \sum_{\tau \in K} \|D^\tau u\|_{L_2(\Omega)}$$

for all α *with* $|\alpha| \leqq 2m$ *and* $0 \leqq \alpha_i \leqq m - 1$ *for all* $1 \leqq i \leqq n$, *where* K *is the set of all indices* $\tau = (\tau_1, \tau_2, \cdots, \tau_n)$ *with* $|\tau| = 2m$ *such that the polynomial* $x^\tau \equiv x_1^{\tau_1} x_2^{\tau_2} \cdots x_n^{\tau_n}$ *is not identically its own mth Hermite interpolation. The result is also true for* $\Omega = R^n$.

Note that the set K of Theorem 3.3 always contains the indices $(2m, 0, \cdots, 0)$, $(0, 2m, 0, \cdots, 0)$, $\cdots$, $(0, 0, \cdots, 2m)$, but for $n > 2$, K contains other indices as well.

The results given thus far can be viewed as results concerning the interpolation and approximation by the tensor product of *one-dimensional* Hermite splines. General results concerning approximation by tensor products of one-dimensional *splines* in higher dimensions have also been established by several authors. In analogy with the notation of § 1.1, let $\text{Sp}^{(m)}(\Delta_i, [a_i, b_i])$ be the collection of splines defined on $[a_i, b_i]$, of local degree $2m - 1$ on subdivisions defined by Δ_i. For notation, let $\Omega \equiv \prod_{i=1}^n (a_i, b_i)$ be a rectangular parallelopiped in R^n, and let $\Delta \equiv \prod_{i=1}^n \Delta_i$, where $\Delta_i \in \mathscr{P}(a_i, b_i)$. Defining $\pi = \max_{1 \leqq i \leqq n} \pi_i$, $\underline{\pi} = \min_{1 \leqq i \leqq n} \underline{\pi}_i$, we say that $\Delta \in \mathscr{P}_\sigma(\Omega)$ if $\pi/\underline{\pi} \leqq \sigma$. Then Schultz [3.7] has proved the following theorem.

THEOREM 3.4. *Given $u \in W_2^r(\Omega)$, where $\Omega = \prod_{i=1}^n (a_i, b_i) \subset R^n$, and given $\Delta = \prod_{i=1}^n \Delta_i \in \mathscr{P}_\sigma(\Omega)$, then there exists a $w \in \bigotimes_{i=1}^n \mathrm{Sp}^{(m)}(\Delta_i, [a_i, b_i])$ such that*

$$(3.1.14) \qquad \|u - w\|_{W_2^p(\Omega)} \leqq K\pi^{r-p}\|u\|_{W_2^r(\Omega)},$$

where $0 \leqq p \leqq \min(r, 2m - 1)$.

Using an approach of Harrick [3.8], suitable extensions can be made (cf. [3.7]) for n-dimensional regions $\Omega \subset \prod_{i=1}^n [a_i, b_i]$ by considering approximations in $\theta(x) \prod_{i=1}^n H^{(m)}(\Delta_i, [a_i, b_i])$, where $\theta(x)$ is positive in Ω and vanishes suitably on $\partial\Omega$ (for details, see [3.7]).

It is interesting that Bramble and Hilbert [3.5], [3.6] also have approximation results, analogous to Theorem 3.3, for *splines*, which are effectively treated as tensor products of one-dimensional splines on a uniform mesh of side h. If $S_h^k, k \geqq 2$, is the collection of all functions u which have continuous partial derivatives of order $k - 2$ in R^n, and, on any hypercube of side h, u is a polynomial of degree $k - 1$ in each variable, then it can be shown (actually via interpolation) that given any $u \in W_2^k(R^n)$,

$$(3.1.15) \qquad \inf_{w \in S_h^k} \|u - w\|_{W_2^j(R^n)} \leqq Ch^{k-j}\|u\|_{W_2^k(R^n)}, \qquad\qquad 0 \leqq j \leqq k.$$

From this, using the Calderón extension theorem as in the proof of Theorem 3.3, one obtains (cf. [3.5], [3.6]) the following.

THEOREM 3.5. *For any $u \in W_2^k(\Omega)$,*

$$(3.1.16) \qquad \inf_{w \in S_h^k} \|u - w\|_{W_2^j(\Omega)} \leqq Ch^{k-j}\|u\|_{W_2^k(\Omega)}, \qquad\qquad 0 \leqq j \leqq k.$$

Based on notions of *quasi-interpolation*, de Boor and Fix [3.9] have obtained deep results which are like those of (3.1.16), but in the norms $W_\infty^j(\Omega)$ (cf. (5.1.20)).

3.2. Zlámal-type extensions. Taking the tensor product to attack higher-dimensional approximation problems for piecewise-polynomial functions is just one way of extending one-dimensional results. Another approach, more closely allied to finite-element methods is due to M. Zlámal [3.10]–[3.12], and can be described as follows.

Assume that Ω is a bounded region in R^2 which can be triangulated, i.e., Ω can be exactly decomposed into a finite number of triangular subregions T_i, $1 \leqq i \leqq N$. Fixing i and calling $T \equiv T_i$, consider any polynomial of degree two in each variable:

$$(3.2.1) \qquad p(x_1, x_2) = \alpha_1 + \alpha_2 x_1 + \alpha_3 x_2 + \alpha_4 x_1^2 + \alpha_5 x_1 x_2 + \alpha_6 x_2^2.$$

If P_i are the vertices of T (see Fig. 1) and Q_i are the midpoints of the sides of T, $1 \leqq i \leqq 3$, then for any constants $\mu_i, \sigma_i, 1 \leqq i \leqq 3$, it is easy to see that there exists a unique $p(x_1, x_2)$ of the form (3.2.1) such that

$$p(P_i) = \mu_i, \quad p(Q_i) = \sigma_i, \qquad\qquad 1 \leqq i \leqq 3.$$

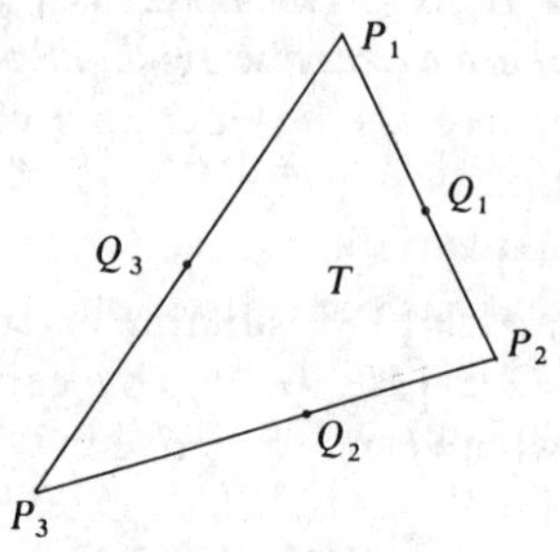

FIG. 1

Thus, if f is a continuous function on T, there exists a unique polynomial $p(x_1, x_2)$ interpolating f on T in the sense that

$$(3.2.2) \qquad f(P_i) = p(P_i), \quad f(Q_i) = p(Q_i), \qquad\qquad 1 \leqq i \leqq 3.$$

Zlámal has shown (cf. [3.10]) the following.

THEOREM 3.6. *Given any $\Omega \subset R^2$ which can be triangulated, i.e., $\bigcup_{i=1}^{N} \overline{T}_i = \overline{\Omega}$, let h denote the largest length of any side of any T_i, and let θ denote the smallest interior angle of any T_i. If $f \in C^3(\overline{\Omega})$, and s is the unique piecewise-polynomial which interpolates f in the sense of (3.2.2), then*

$$(3.2.3) \qquad \| f - s \|_{W_2^1(\Omega)} \leqq K \frac{h^2}{\sin \theta_0} \{ \sup \{ \| D^\alpha f \|_{L_\infty(\overline{\Omega})} : |\alpha| = 3 \} \}$$

for any triangulation with $\theta \geqq \theta_0 > 0$, where K is independent of f and the geometry.

Similar results have been obtained by Zlámal for piecewise-polynomial (of degree 3) interpolations, defined by interpolating f and its two first partial derivatives at each vertex and interpolating f at the center of gravity of each T_i.

Since an interval and a triangle are n-simplices for $n = 1$ and $n = 2$, respectively, it would seem natural to generalize Zlámal's results to an n-dimensional setting via n-simplices (an n-simplex is the convex hull of $n + 1$ noncoplanar points in R^n), as done recently by Ciarlet and Wagschal [3.13]. The inherent shortcomings of these results of [3.10]–[3.13] can, however, be seen by the typical result of Theorem 3.6 above. As in Theorem 3.1 and its Corollary 3.2, the function space setting for the error bound of (3.2.3) is again not what one would naturally expect; one would expect h^2 accuracy for $f \in W_2^3(\Omega)$ in (3.2.3). Fortunately, Zlámal and Bramble [3.14] have established an improved and generalized version of Theorem 3.6, which we now describe.

Given a bounded region Ω in R^n, whose boundary $\partial\Omega$ is a simplicial complex (a generalization to R^n of a polygon in R^2), assume a generalized triangulation T over $\overline{\Omega}$, i.e., $\overline{\Omega}$ is the set-theoretic union of a finite number of n-simplices S_i, $1 \leqq i \leqq N$, whose interiors are pair-wise disjoint and such that, given any n-simplex S_i of the triangulation, each one of its $(n-1)$-faces is either a portion of the boundary $\partial\Omega$ or else is also an $(n-1)$-face of another n-simplex of the triangulation. For ease of description here, assume that all simplices S_i are *equilateral* and that the region Ω can be *triangulated* by such regular simplices S_i^h for a sequence of $h \to 0$, where h is the length of a common edge of an equilateral simplex. Next, if $T_h(R^n)$ is the subspace of $C^0(R^n)$ of all functions which are polynomials *of degree two* in each variable on each (regular) simplex S_i^h of $R^n = \bigcup_i S_i^h$ of edge h, then $T_h(\Omega)$ is defined as the restriction of $T_h(R^n)$ to Ω. Given $u \in C^0(\Omega)$, there is, in the manner of (3.2.2), a unique $w_h \in T_h(\Omega)$ which *interpolates* u at each vertex and midpoint of each edge of the triangulation of Ω. Then, in the manner of (3.1.8)–(3.1.12), one obtains the following theorem (cf. [3.14]).

THEOREM 3.7. *Let Ω be a bounded region in R^n which can be triangulated by regular simplices S_i^h for a sequence of $h \to 0$. Then, for $u \in W_2^3(\Omega)$,*

$$(3.2.4) \qquad \inf_{w \in T_h(\Omega)} \|u - w\|_{W_2^j(\Omega)} \leqq Ch^{3-j} \|u\|_{W_2^3(\Omega)}, \qquad j = 0, 1,$$

where C is independent of h and u.

It is important to note that the results of Theorems 3.3 and 3.7, while proved either for hypercubes or regular simplices in R^n, do extend to more general regions. In the case of Theorem 3.3, rectangular parallelepipeds may be used, provided that the ratio of the lengths of any edges remains bounded above and below for any h. The same is true of Theorem 3.7.

For computer implementation of these so-called Zlámal-type finite element methods, see George [3.15].

REFERENCES

[3.1] G. BIRKHOFF, M. H. SCHULTZ AND R. S. VARGA, *Piecewise Hermite interpolation in one and two variables with applications to partial differential equations*, Numer. Math., 11 (1968), pp. 232–256.

[3.2] A. SARD, *Linear Approximation*, Math. Survey 9, American Mathematical Society, Providence, Rhode Island, 1963.

[3.3] S. AGMON, *Lectures on Elliptic Boundary Value Problems*, Van Nostrand, Princeton, New Jersey, 1965.

[3.4] J. H. BRAMBLE AND S. R. HILBERT, *Bounds for a class of linear functionals with applications to Hermite interpolation*, Numer. Math., 16 (1971), pp. 362–369.

[3.5] ———, *Estimation of linear functionals on Sobolev spaces with applications to Fourier transforms*, SIAM J. Numer. Anal., 7 (1970), pp. 112–124.

[3.6] S. R. HILBERT, *Numerical methods for elliptic boundary value problems*, Thesis, University of Maryland, 1969.

[3.7] M. H. SCHULTZ, *Multivariate spline functions and elliptic problems*, Approximations with Special Emphasis on Spline Functions, I. J. Schoenberg, ed., Academic Press, New York, 1969, pp. 279–347.

[3.8] I. I. HARRICK, *Approximation of functions which vanish on the boundary of a region, together with their partial derivatives, by functions of special type*, Akad. Nauk. SSSR Izv. Sibirsk. Otd., 4 (1963), pp. 408–425.

[3.9] C. DE BOOR AND G. FIX, *Spline approximation by quasi-interpolants*, J. Approx. Theory, to appear.

[3.10] M. ZLÁMAL, *On the finite element method*, Numer. Math., 12 (1968), pp. 394–409.

[3.11] ———, *On some finite element procedures for solving second order boundary value problems*, Ibid., 14 (1969), pp. 42–48.

[3.12] ——, *A finite element procedure of the second order accuracy*, Ibid., 14 (1970), pp. 394–402.

[3.13] P. G. CIARLET AND C. WAGSCHAL, *Multipoint Taylor formulas and applications to the finite element method*, Ibid., 17 (1971), pp. 84–100.

[3.14] J. BRAMBLE AND M. ZLÁMAL, *Triangular elements in the finite element method*, Math. Comp., 24 (1970), pp. 809–820.

[3.15] J. A. GEORGE, *Computer implementation of the finite element method*, Thesis, Rep. CS208, Computer Science Department, Stanford University, California, 1971.

CHAPTER 4

The Rayleigh–Ritz–Galerkin Method for Nonlinear Boundary Value Problems

4.1. One-dimensional problem. To show how theorems about interpolation and approximation by piecewise-polynomial functions can be used to deduce results about approximate solutions of nonlinear boundary value problems, we first discuss two-point boundary value problems, as thoroughly considered in Keller [4.1].

Specifically, we shall consider problems of the form

$$(4.1.1) \qquad \mathscr{L}u(x) = f(x, u(x)), \qquad -\infty < a < x < b < \infty,$$

with homogeneous Dirichlet boundary conditions

$$(4.1.2) \qquad D^k u(a) = D^k u(b) = 0, \qquad 0 \le k \le m - 1,$$

where the differential operator $\mathscr{L}$ in self-adjoint form is given by

$$(4.1.3) \qquad \mathscr{L}u(x) \equiv \sum_{j=0}^{m} (-1)^j D^j \{p_j(x) D^j u(x)\}, \qquad m \ge 1.$$

For one-dimensional problems, *nonhomogeneous* boundary conditions $D^k u(a) = \alpha_k$, $D^k u(b) = \beta_k$, $0 \le k \le m - 1$, can always be reduced to the form (4.1.2) by means of a suitable change of dependent variable. Other types of boundary conditions, such as nonlinear, Neumann, and mixed boundary conditions in one dimension can also be treated (cf. [4.2], [4.3] and [4.4]).

For specific assumptions about $\mathscr{L}$, we assume that all p_j are bounded on $[a, b]$, $0 \le j \le m$, and that the operator $\mathscr{L}$ of (4.1.3) is $\mathring{W}_2^m[a, b]$-*elliptic*, i.e., there exists a positive constant K such that

$$(4.1.4) \quad K\|w\|^2_{\mathring{W}_2^m[a,b]} \le \int_a^b \left\{ \sum_{j=0}^{m} p_j(x)(D^j w(x))^2 \right\} dx \quad \text{for all} \quad w \in \mathring{W}_2^m[a, b],$$

where we recall that $\mathring{W}_2^m[a, b]$ is the collection of all real-valued functions $w(x)$, where $D^{m-1} w(x)$ is absolutely continuous on $[a, b]$, $D^m w \in L_2[a, b]$, and where $D^j w(a) = D^j w(b) = 0$ for $0 \le j \le m - 1$. In addition, we assume that $f(x, u)$ is a real-valued measurable function on $[a, b] \times R$ such that $f(x, v(x)) \in L_2[a, b]$ for all $v \in \mathring{W}_2^m[a, b]$, and such that f satisfies

$$(4.1.5) \qquad \frac{f(x, u) - f(x, v)}{u - v} \le \gamma < \Lambda \equiv \inf_{\substack{w \in \mathring{W}_2^m[a,b] \\ w \not\equiv 0}} \frac{\displaystyle\int_a^b \left\{ \sum_{j=0}^{m} p_j (D^j w)^2 \right\} dx}{\displaystyle\int_a^b w^2(x)\, dx}$$

for almost all $x \in [a, b]$ and all $-\infty < u, v < \infty$ with $u \neq v$. Note that since the denominator of the last term of (4.1.5) is bounded above from (1.1.3) by $\|w\|^2_{\mathring{W}^m_2[a,b]}$, it then follows from the $\mathring{W}^m_2[a, b]$-ellipticity assumption of (4.1.4) that Λ is positive.

We next assume that for each constant $c > 0$, there exists a positive constant $M(c)$ such that

$$(4.1.6) \qquad \left| \frac{f(x, u) - f(x, v)}{u - v} \right| \leqq M(c) \quad \text{for almost all} \quad x \in [a, b] \quad \text{and all} \quad u \neq v$$

$$\text{with} \quad |u| \leqq c \quad \text{and} \quad |v| \leqq c.$$

Defining the *quasi-bilinear form*

$$(4.1.7) \qquad B(u, v) \equiv \int_a^b \left\{ \sum_{j=0}^m p_j(x) D^j u(x) D^j v(x) \right\} dx - \int_a^b f(x, u(x)) \cdot v(x) \, dx$$

for all $u, v \in \mathring{W}^m_2[a, b]$, we say that the boundary value problem of (4.1.1)–(4.1.2) admits a *generalized* solution u in $\mathring{W}^m_2[a, b]$ (cf. Browder [4.5]) if

$$(4.1.8) \qquad\qquad B(u, v) = 0 \quad \text{for all} \quad v \in \mathring{W}^m_2[a, b].$$

A general result, based on the theory of monotone operators to be discussed in § 4.2, is the following theorem.

THEOREM 4.1. *With the assumptions of* (4.1.4)–(4.1.6), *then the nonlinear boundary value problem* (4.1.1)–(4.1.2) *admits a unique generalized solution in* $\mathring{W}^m_2[a, b]$.

Next, if S_M is any finite-dimensional subspace of $\mathring{W}^m_2[a, b]$, then, in analogy with (4.1.8), we would call w_M in S_M the *Galerkin approximation* of the generalized solution u of (4.1.1)–(4.1.2) if

$$(4.1.9) \qquad\qquad B(w_M, v) = 0 \quad \text{for all} \quad v \in S_M.$$

The next result shows that w_M, so defined, is uniquely determined, and gives error bounds for $u - w_M$.

THEOREM 4.2. *With the assumptions of* (4.1.4)–(4.1.6), *there is a unique w_M in S_M which satisfies* (4.1.9). *Moreover, there exist constants K and K', independent of the choice of S_M, such that*

$$(4.1.10) \quad \|D^j(u - w_M)\|_{L_\infty[a,b]} \leqq K \|u - w_M\|_{\mathring{W}^m_2[a,b]} \leqq K' \inf_{w \in S_M} \|u - w\|_{\mathring{W}^m_2[a,b]}$$

for all $0 \leqq j \leqq m - 1$, where u is the unique generalized solution of (4.1.1)–(4.1.2) *in* $\mathring{W}^m_2[a, b]$.

We shall show in § 4.2 that the second inequality of (4.1.10) is a consequence of Theorem 4.6 on monotone operators. The first inequality of (4.1.10) is elementary to establish, and we give a direct proof. For any $v \in \mathring{W}^m_2[a, b]$ and any nonnegative integer j with $0 \leq j \leq m - 1$, the fact that $D^j v(a) = D^j v(b) = 0$ allows us to write that

$$D^j v(x) = \int_a^x D^{j+1} v(t)\, dt = -\int_x^b D^{j+1} v(t)\, dt \quad \text{for any} \quad x \in [a, b].$$

Hence,

$$(4.1.11) \qquad 2 D^j v(x) = \int_a^b \operatorname{sgn}(x - t) D^{j+1} v(t)\, dt,$$

where $\operatorname{sgn} y \equiv 1$ for any $y \geq 0$ and $\operatorname{sgn} y \equiv -1$ if $y < 0$. Thus,

$$2 \| D^j v \|_{L_\infty[a,b]} \leq (b - a) \| D^{j+1} v \|_{L_\infty[a,b]} \quad \text{for} \quad j = 0, 1, \cdots, m - 2,$$

while for $j = m - 1$, the application of Schwarz's inequality to (4.1.11) yields

$$2 \| D^{m-1} v \|_{L_\infty[a,b]} \leq (b - a)^{1/2} \| D^m v \|_{L_2[a,b]} \leq (b - a)^{1/2} \| v \|_{W_2^m[a,b]}.$$

It is then clear from the above two inequalities that there exists a positive constant K for which

$$\| D^j v \|_{L_\infty[a,b]} \leq K \| v \|_{W_2^m[a,b]} \quad \text{for all} \quad 0 \leq j \leq m - 1,$$

which is the first inequality of (4.1.10). The above inequality is just a special case of the *Sobolev Imbedding Theorem* in one-spatial variable (cf. Yosida [4.6, p. 174]).

It is now an easy matter to apply the interpolation and approximation errors for piecewise-polynomial subspaces of $\mathring{W}_2^m[a, b]$. Consider any L-spline space $\mathrm{Sp}(M, \Delta, z)$, where M is of order r, $r \geq 1$, and the positive integer z is such that $1 \leq z \leq r$. Then, it follows that $\mathrm{Sp}(M, \Delta, z) \subset W_2^{2r-z}[a, b]$. If we assume that $2r - z \geq m$, and if we restrict attention to the subspace $\mathrm{Sp}_0(M, \Delta, z)$ of $\mathrm{Sp}(M, \Delta, z)$ whose elements satisfy the boundary conditions of (4.1.2), then $\mathrm{Sp}_0(M, \Delta, z)$ is a finite-dimensional subspace of $\mathring{W}_2^m[a, b]$. As such, we can couple the results of Theorems 4.1 and 4.2 with the error bounds of (1.2.20) of Corollary 1.10.

THEOREM 4.3. *With the assumptions of* (4.1.4)–(4.1.6), *let* u *be the unique generalized solution of* (4.1.1)–(4.1.2) *in* $\mathring{W}_2^m[a, b]$. *If* $S_M \equiv \mathrm{Sp}_0(M, \Delta, z) \subset \mathring{W}_2^m[a, b]$, *where* M *is of order* r *and* $2r - z \geq m$, *let* w_M *be the unique element in* S_M *which satisfies* (4.1.9). *If* $u \in W_2^\sigma[a, b]$, *where* $m < \sigma \leq 2r$, *then*

$$(4.1.12) \qquad \| D^j(u - w_M) \|_{L_\infty[a,b]} \leq K \| u - u_M \|_{W_2^m[a,b]} \leq K \pi^{\sigma - m}, \quad 0 \leq j \leq m - 1.$$

Of course, related results can be stated for Lg-spline and γ-spline subspaces of $\mathring{W}_2^m[a, b]$ as well.

It is not difficult to show that the error bounds of (4.1.12) in the norm $\| \cdot \|_{W_2^m[a,b]}$ are in general *best possible*. But, the corresponding error bounds in the norm $\| \cdot \|_{L_\infty[a,b]}$ are *not*, basically because they were derived as consequences of error bounds in $\| \cdot \|_{W_2^m[a,b]}$. To be more specific, consider the following special case of (4.1.1)–(4.1.2):

$$(4.1.13) \qquad \begin{aligned} -D^2 u(x) &= f(x, u), & 0 < x < 1, \\ u(0) &= u(1) = 0, \end{aligned}$$

where $f(x, u) \in C^0([0, 1] \times R)$, and where f satisfies the hypotheses of (4.1.5)–(4.1.6), with $\Lambda = \pi^2$. Using the Hermite subspace $H_0^{(1)}(\Delta_u) \subset \mathring{W}_2^1[0, 1]$, it follows from Theorem 4.3 that if the unique generalized solution u is an element of $C^2[0, 1]$, then from (4.1.12) (with $\sigma = 2$, $m = 1$),

$$\|u - w_M\|_{L_\infty[0,1]} \leqq K\pi.$$

Ciarlet [4.7] improved the above error bound to

$$\|u - w_M\|_{L_\infty[0,1]} \leqq K\pi^2,$$

and this has subsequently been generalized in Perrin, Price and Varga [4.8]. We sketch these developments below.

Given the nonlinear boundary value problem of (4.1.1)–(4.1.2), choose the specific γ-spline subspace $S_0(\mathscr{L}, \Delta, z)$, where $\mathscr{L}$ is given by (4.1.3). In this development, it is important that the differential operator $\mathscr{L}$ of (4.1.3) be chosen equal to the operator E of (2.2.1) defining the γ-spline space. Then, if u is the generalized solution of (4.1.1)–(4.1.2) in $\mathring{W}_2^m[a, b]$, and if w_M is its approximation in $S_0(\mathscr{L}, \Delta, z)$ (cf. Theorem 4.3), let $\tilde{w}$ be the interpolation of u in $S_0(\mathscr{L}, \Delta, z)$, in the sense of (2.2.4). Following the construction of [4.8], it can be shown that

(4.1.14) $$\|\tilde{w} - w_M\|_{W_2^m[a,b]} \leqq K\|\tilde{w} - u\|_{L_2[a,b]}.$$

If $u \in W_2^\sigma[a, b]$, where $m \leqq \sigma \leqq 2m$, it follows from Theorem 2.3 and Corollary 1.10 that

(4.1.14′) $$\|D^j(\tilde{w} - u)\|_{L_2[a,b]} \leqq K\pi^{\sigma-j}, \qquad 0 \leqq j \leqq m,$$

so that the inequality of (4.1.14), using (4.1.14′) with $j = 0$, reduces to

(4.1.14″) $$\|\tilde{w} - w_M\|_{W_2^m[a,b]} \leqq K\pi^\sigma.$$

Hence, from the triangle inequality and the inequalities of (4.1.14′) and (4.1.14″),

$$\|D^j(u - w_M)\|_{L_2[a,b]} \leqq \|D^j(u - \tilde{w})\|_{L_2[a,b]} + \|D^j(\tilde{w} - w_M)\|_{L_2[a,b]}$$

$$\leqq K\pi^{\sigma-j} + \|\tilde{w} - w_M\|_{W_2^m[a,b]}$$

$$\leqq K\pi^{\sigma-j} + K\pi^\sigma \leqq K\pi^{\sigma-j}$$

for any $0 \leqq j \leqq m$. Similar results in $\|\cdot\|_{L_\infty[a,b]}$ can also be established, and are stated in the following theorem.

THEOREM 4.4. *With the assumptions of (4.1.4)–(4.1.6), let u be the unique generalized solution of (4.1.1)–(4.1.2) in $\mathring{W}_2^m[a,b]$, and assume further that $u \in W_2^\sigma[a,b]$, where $m \leqq \sigma \leqq 2m$. If $S_M \equiv S_0(\mathscr{L}, \Delta, z)$ and w_M is the unique element in S_M which satisfies (4.1.9), then*

(4.1.15) $$\|D^j(u - w_M)\|_{L_2[a,b]} \leqq K\pi^{\sigma-j}, \qquad 0 \leqq j \leqq m.$$

Furthermore, if $u \in C^\sigma[a, b]$, $m \leqq \sigma \leqq 2m$, and if $\tilde{w}$, the unique interpolation of u in $S_0(\mathscr{L}, \Delta, z)$ in the sense of (2.2.4), satisfies

(4.1.16) $$\|D^j(u - \tilde{w})\|_{L_\infty[a,b]} \leqq K\pi^{\sigma-j}, \qquad 0 \leqq j \leqq m - 1,$$

then

$$(4.1.17) \qquad \|D^j(u - w_M)\|_{L_\infty[a,b]} \leqq K\pi^{\sigma-j}, \qquad 0 \leqq j \leqq m - 1.$$

We remark that Hermite L-splines and polynomial splines (in $\mathrm{Sp}^{(m)}(\Delta)$), defined for a uniform partition Δ_u of $[a, b]$, satisfy the required interpolation accuracy of (4.1.16) (cf. Swartz and Varga [4.9]), so that there are many cases in which the improved uniform estimates of (4.1.17) are in fact valid.

It is also interesting to note that higher order accuracies can be obtained, using a construction of Hulme [4.10], which generalized a result of Rose [4.11] (for details, see also [4.8]).

4.2. Monotone operator theory. In this section, we discuss briefly the theory of *monotone operators*, due to Zarantonello [4.12], Browder [4.5] and Minty [4.13]. For a more complete treatment stressing applications, see [4.14]. In the next section, this will be used in the discussion of Galerkin approximations of solutions of *nonlinear* elliptic boundary value problems.

Let H be a real Hilbert space with inner product $(\cdot, \cdot)_H$, and norm $\|\cdot\|$, and let T be a (possibly nonlinear) mapping from H into H satisfying the following hypotheses:

(4.2.1) T is *finitely continuous*, i.e., T is continuous from finite-dimensional subspaces of H into H with the weak-star topology. In other words, given any finite-dimensional subspace H_k of H and any sequence $\{u_n\}_{n=1}^\infty$ of elements of H_k which converges to an element u in H, then the sequence $\{(Tu_n, v)_H\}_{n=1}^\infty$ converges to $(Tu, v)_H$ for any $v \in H$;

(4.2.2) T is *strongly monotone*, i.e., there exists a positive constant K for which $K\|u - v\|^2 \leqq (Tu - Tv, u - v)_H$ for all $u, v \in H$.

Weaker conditions, formulated in terms of reflexive Banach spaces, can be found in Browder [4.5] and [4.14].

Consider the problem of determining $u \in H$ such that

$$(4.2.3) \qquad\qquad Tu = 0,$$

or equivalently,

$$(4.2.4) \qquad\qquad (Tu, v)_H = 0 \quad \text{for all} \quad v \in H.$$

The abstract Galerkin method corresponding to $Tu = 0$ in (4.2.3) consists of finding a u_k in H_k, where H_k is any finite-dimensional subspace of H, which analogously satisfies

$$(4.2.5) \qquad\qquad (Tu_k, v)_H = 0 \quad \text{for all} \quad v \in H_k.$$

The next result is due to Browder [4.5].

THEOREM 4.5. *Let T be finitely continuous and strongly monotone. Then, $Tu = 0$ in (4.2.3) has a unique solution u. Similarly, for any finite-dimensional subspace H_k of H, there exists a unique u_k in H_k, the Galerkin approximation of u in H_k, which satisfies (4.2.5).*

To study the convergence of the Galerkin approximation u_k in H_k to the solution u of $Tu = 0$, we need the additional assumption that T is *bounded*, i.e., T maps bounded subsets of H into bounded subsets of H (cf. [4.14]).

THEOREM 4.6. *Let T be finitely continuous and strongly monotone, and assume that T is bounded. If u is the unique solution of $Tu = 0$ and u_k is its unique Galerkin approximation in H_k (cf. (4.2.5)), then there exists a positive constant K such that*

$$(4.2.6) \qquad \|u - u_k\|^2 \leqq K \inf_{w \in H_k} \|u - w\|$$

for any finite-dimensional subspace H_k of H. Moreover, if T is Lipschitz continuous for bounded arguments, i.e., given $M > 0$, there exists a constant $K(M)$ such that

$$(4.2.7) \quad \|Tu - Tv\| \leqq K(M)\|u - v\| \quad \text{for all} \quad u, v \in H \quad \text{with} \quad \|u\|, \|v\| \leqq M,$$

then there exists a positive constant K such that

$$(4.2.8) \qquad \|u - u_k\| \leqq K \inf_{w \in H_k} \|u - w\|$$

for all finite-dimensional subspaces H_k of H.

The whole point of our discussion on monotone operators rests in the inequalities of (4.2.6) and (4.2.8). With suitable choices of the finite-dimensional subspaces H_k of H, we can apply the interpolation and approximation results of Chapters 1–3 directly to

$$\inf_{w \in H_k} \|u - w\|,$$

and this then gives from (4.2.6) and (4.2.8) *upper bounds for the Galerkin error* $\|u - u_k\|$.

As an immediate consequence of Theorem 4.6, we have a corollary.

COROLLARY 4.7. *Let T be a finitely continuous, bounded, and strongly monotone mapping of H into itself, and let $\{H_k\}_{k=1}^{\infty}$ be a sequence of finite-dimensional subspaces of H such that*

$$(4.2.9) \qquad \lim_{k \to \infty} \left\{ \inf_{w \in H_k} \|u - w\| \right\} = 0,$$

where u is the unique solution of (4.2.3). Then,

$$(4.2.10) \qquad \lim_{k \to \infty} \|u - u_k\| = 0,$$

where $u_k, k = 1, 2, \cdots$, is the unique Galerkin approximation of u in H_k.

4.3. Galerkin approximation of nonlinear boundary value problems. In this section, we consider applications of the theory of monotone operators to the approximate solution of nonlinear boundary value problems.

Let Ω be a bounded region in R^n, $n \geqq 1$, whose boundary $\partial\Omega$ is sufficiently smooth; for convenience, assume that Ω satisfies a restricted cone condition, as described in § 3.1. We then consider the $2m$th order Dirichlet problem:

$$(4.3.1) \quad \sum_{|\alpha| \le m} (-1)^{|\alpha|} D^\alpha \{ A_\alpha(x, u(x), \cdots, D^m u(x)) \} = 0, \qquad x \in \Omega,$$

$$D^\beta u(x) = 0, \qquad x \in \partial\Omega \quad \text{for all } 0 \le |\beta| \le m - 1,$$

where $A_\alpha(x, u, \cdots, D^m u)$ denotes a function which can depend upon x and *any* $D^\gamma u$ with $|\gamma| \le m$. With the notation of § 3.1 we recall that the Hilbert space $\mathring{W}_2^m(\Omega)$ is defined as the completion of all real functions f which are infinitely differentiable in Ω and have compact support in Ω (written $f \in C_0^\infty(\Omega)$) with respect to the norm

$$(4.3.2) \quad \| f \|_{W_2^m(\Omega)}^2 \equiv \sum_{|\alpha| \le m} \int_\Omega (D^\alpha f)^2 \, dx.$$

For any $u, v \in \mathring{W}_2^m(\Omega)$, we formally define the *quasi-bilinear* form

$$(4.3.3) \quad B(u, v) \equiv \sum_{|\alpha| \le m} \int_\Omega A_\alpha(x, u, \cdots, D^m u) \cdot D^\alpha v \, dx,$$

and we say that (4.3.1) admits a generalized solution u in $\mathring{W}_2^m(\Omega)$ if

$$(4.3.4) \quad B(u, v) = 0 \quad \text{for all} \quad v \in \mathring{W}_2^m(\Omega).$$

Formally, one obtains the quasi-bilinear form (4.3.3) and the expression (4.3.4), as follows. Assuming u to be a solution of (4.3.1), multiply the first equation of (4.3.1) by an arbitrary $v \in \mathring{W}_2^m(\Omega)$ and integrate the resultant product over Ω. With integration by parts and the particular boundary conditions of (4.3.1), then (4.3.4) follows.

From [4.14, Theorem 3.1], we state the following.

THEOREM 4.8. *Let the functions* $A_\alpha(x, s_1, \cdots, s_m)$ *appearing in* (4.3.1) *be measurable in* $x \in \Omega$ *and continuous in their other arguments* s_j *for almost all* $x \in \Omega$. *Let* $g(r)$ *be a nonnegative continuous function on* $[0, +\infty)$ *such that for any* $u \in \mathring{W}_2^m(\Omega)$,

$$(4.3.5) \quad \begin{aligned} &|A_\alpha(x, u, \cdots, D^m u)| \\ &\le g \left\{ \sum_{|\beta| < m - n/2} |D^\beta u| \right\} \cdot \left\{ 1 + \sum_{|\beta| = m - n/2} |D^\beta u| + \sum_{m - n/2 < |\beta| \le m} |D^\beta u|^{q_\beta / p_\alpha} \right\} \end{aligned}$$

for all $|\alpha| \le m$, *almost all* $x \in \Omega$, *and all* $D^\gamma u, |\gamma| \le m$, *where*

$$q_\beta = 2n/(n - 2m + 2|\beta|)$$

and

$$\frac{1}{p_\alpha} = \begin{cases} 1 & \text{if} \quad |\alpha| < m - \dfrac{n}{2}, \\[2ex] \dfrac{1}{2} + \dfrac{m - |\alpha|}{n} & \text{if} \quad |\alpha| \ge m - \dfrac{n}{2}. \end{cases}$$

Then, for any $u \in \mathring{W}_2^m(\Omega)$, *there exists a constant* $K = K_u$, *depending on* u, *such that*

$$(4.3.6) \quad |B(u, v)| \le K_u \cdot \| v \|_{W_2^m(\Omega)} \quad \text{for all} \quad v \in \mathring{W}_2^m(\Omega).$$

We remark that the proof of Theorem 4.8 uses the *Sobolev Imbedding Theorem* (cf. [4.6, p. 174]). It is here that one needs a sufficiently smooth boundary $\partial\Omega$, so that the Sobolev Imbedding Theorem is valid.

As a consequence of (4.3.6), the quasi-bilinear form of (4.3.3) is, for each $u \in \mathring{W}_2^m(\Omega)$, a bounded linear functional in v on $\mathring{W}_2^m(\Omega)$. As such, the Riesz representation theorem (cf. [4.6, p. 90]) gives us that there is a unique $Tu \in \mathring{W}_2^m(\Omega)$ such that

$$(4.3.7) \qquad B(u, v) = (Tu, v)_{W_2^m(\Omega)} \quad \text{for all} \quad v \in \mathring{W}_2^m(\Omega),$$

where $(\,\cdot\,,\cdot\,)_{W_2^m(\Omega)}$ denotes the inner product on $\mathring{W}_2^m(\Omega)$. This then defines the mapping T from $\mathring{W}_2^m(\Omega)$ into $\mathring{W}_2^m(\Omega)$.

For properties of the mapping T, we have from [4.13] the following theorem.

THEOREM 4.9. *With the assumptions of Theorem 4.8, the mapping* $T: \mathring{W}_2^m(\Omega)$ $\to \mathring{W}_2^m(\Omega)$, *as given in* (4.3.7), *is bounded and finitely continuous. If moreover, there exists a positive constant K such that*

$$(4.3.8) \quad B(u, u - v) - B(v, u - v) \geqq K \|u - v\|^2_{W_2^m(\Omega)} \quad \text{for all} \quad u, v \in \mathring{W}_2^m(\Omega),$$

then T so defined is strongly monotone. Consequently, the generalized Dirichlet problem of (4.3.1) *admits a unique generalized solution u in* $\mathring{W}_2^m(\Omega)$, *the Galerkin method yields a unique solution u_k on each finite-dimensional subspace H_k of* $\mathring{W}_2^m(\Omega)$, *and there exists a positive constant K such that*

$$(4.3.9) \qquad \|u - u_k\|^2_{W_2^m(\Omega)} \leqq K \inf_{w \in H_k} \|u - w\|_{W_2^m(\Omega)}.$$

The restrictions on the coefficients $A_\alpha(x, u, \cdots, D^m u)$ of (4.3.5) are, of course, very restrictive. Yet, with a priori bounds for the generalized solution u of (4.3.1), a new problem (4.3.1) can be defined so that the new coefficients $\tilde{A}_\alpha(x, u, \cdots, D^m u)$ do satisfy the growth restrictions of (4.3.5). Furthermore, in most interesting cases, the mapping T defined by (4.3.7) is in addition *Lipschitz continuous*, so that one has, in place of (4.3.9), the improved inequality (cf. (4.2.8) of Theorem 4.6)

$$(4.3.10) \qquad \|u - u_k\|_{W_2^m(\Omega)} \leqq K \inf_{w \in H_k} \|u - w\|_{W_2^m(\Omega)}.$$

(For details, see [4.14].) We note that the bounds of (4.3.9) and (4.3.10) do *not* directly give rates of convergence for Galerkin approximations, since this depends on the choices of H_k. Nonetheless, (4.3.9) and (4.3.10) *set the stage* for use of the approximation results of Chapters 1–3.

It is instructive to reexamine the particular nonlinear two-point boundary value problem of (4.1.1)–(4.1.3), in the light of monotone operator theory. The general quasi-bilinear form of (4.3.3) in this case reduces to (cf. (4.1.7))

$$(4.3.11) \qquad B(u, v) \equiv \int_a^b \left\{ \sum_{j=0}^m p_j D^j u \cdot D^j v \right\} dx - \int_a^b f(x, u(x)) \cdot v(x)\, dx,$$

where $u, v \in \mathring{W}_2^m[a, b]$. Since it was assumed in §4.1 that the coefficients $p_j(x)$ are all bounded on $[a, b]$, and that $f(x, u) \in L_2[a, b]$ for all $u \in \mathring{W}_2^m[a, b]$, then it easily follows from Schwarz's inequality applied to the integrals of (4.3.11) that

$$(4.3.11') \qquad |B(u, v)| \leq K_u \|v\|_{W_2^m[a,b]} \quad \text{for all} \quad u, v \in \mathring{W}_2^m[a, b].$$

This, as in (4.3.6), implies that a mapping $T : \mathring{W}_2^m[a, b] \to \mathring{W}_2^m[a, b]$ exists for which

$$(4.3.12) \qquad B(u, v) = (Tu, v)_{W_2^m[a,b]} \quad \text{for all} \quad u, v \in \mathring{W}_2^m[a, b],$$

and, as in Theorem 4.9, this operator T, abstractly defined, is *bounded* and *finitely continuous*. To see if T is strongly monotone, it suffices to show that

$$(4.3.13) \qquad B(u, u - v) - B(v, u - v) \geq K \|u - v\|_{W_2^m[a,b]}^2 \quad \text{for all} \quad u, v \in \mathring{W}_2^m[a, b].$$

From (4.3.11), we see that

$$B(u, u - v) - B(v, u - v) = \sum_{j=0}^{m} \int_a^b p_j(D^j(u - v))^2 \, dx$$
$$- \int_a^b \left[\frac{f(x, u) - f(x, v)}{u - v} \right](u - v)^2 \, dx.$$

With the hypothesis of (4.1.5), this becomes

$$B(u, u - v) - B(v, u - v) \geq \left(\frac{\Lambda - \gamma}{\Lambda} \right) \sum_{j=0}^{m} \int_a^b p_j(D^j(u - v))^2 \, dx,$$

and, with the assumption of (4.1.4), the above inequality reduces to the desired inequality of (4.3.13), i.e., T is *strongly monotone*.

Finally, it remains to show that T, as defined in (4.3.12), is Lipschitz continuous (cf. (4.2.7)). Because the norm of $Tu - Tv$ can also be expressed as

$$(4.3.14) \qquad \|Tu - Tv\|_{W_2^m[a,b]} \equiv \sup_{w \not\equiv 0} \frac{|(Tu - Tv, w)_{W_2^m[a,b]}|}{\|w\|_{W_2^m[a,b]}},$$

it suffices to consider then $(Tu - Tv, w)_{W_2^m[a,b]}$, or equivalently from (4.3.12), $B(u, w) - B(v, w)$. But, with the explicit assumption of (4.1.6), it is easily verified that

$$(4.3.15) \qquad |(Tu - Tv, w)_{W_2^m[a,b]}| \leq K(M)\|u - v\|_{W_2^m[a,b]} \cdot \|w\|_{W_2^m[a,b]}$$

for all $u, v, w \in W_2^m[a, b]$ with $\|u\|_{W_2^m[a,b]}, \|v\|_{W_2^m[a,b]} \leq M$. But then, with (4.3.14), the inequality of (4.3.15) gives us that

$$(4.3.16) \qquad \|Tu - Tv\|_{W_2^m[a,b]} \leq K(M)\|u - v\|_{W_2^m[a,b]},$$

i.e., T is *Lipschitz continuous for bounded arguments*.

REFERENCES

[4.1] H. B. KELLER, *Numerical Methods for Two-Point Boundary-Value Problems*, Blaisdell, Waltham, Mass., 1968.

[4.2] P. G. CIARLET, M. H. SCHULTZ AND R. S. VARGA, *Numerical methods of high-order accuracy for nonlinear boundary value problems. I. One-dimensional problem*, Numer. Math., 9 (1967), pp. 394–430.

[4.3] ———, *II. Nonlinear boundary conditions*, Ibid., 11 (1968), pp. 331–345.

[4.4] ———, *IV. Periodic boundary conditions*, Ibid., 12 (1968), pp. 266–279.

[4.5] F. E. BROWDER, *Existence and uniqueness theorems for solutions of nonlinear boundary value problems*, Proc. Amer. Math. Soc. Symposia in Appl. Math., 17 (1965), pp. 24–49.

[4.6] K. YOSIDA, *Functional Analysis*, Academic Press, New York, 1965.

[4.7] P. G. CIARLET, *An $O(h^2)$ method for a non-smooth boundary value problem*, Aequationes Mathematicae, 2 (1968), pp. 39–49.

[4.8] F. M. PERRIN, H. S. PRICE AND R. S. VARGA, *On higher order methods for nonlinear two-point boundary value problems*, Numer. Math., 13 (1969), pp. 180–198.

[4.9] B. SWARTZ AND R. S. VARGA, *Error bounds for spline and L-spline interpolation*, J. Approx. Theory, to appear.

[4.10] B. L. HULME, *Interpolation by Ritz approximation*, J. Math. Mech., 18 (1968), pp. 337–342.

[4.11] M. E. ROSE, *Finite difference schemes for differential equations*, Math. Comp., 18 (1964), pp. 179–195.

[4.12] E. H. ZARANTONELLO, *Solving functional equations by contractive averaging*, MRC Tech. Rep. 160, Mathematics Research Center, United States Army, University of Wisconsin, Madison, 1960.

[4.13] G. MINTY, *Monotone (nonlinear) operators in Hilbert space*, Duke Math. J., 29 (1962), pp. 341–346.

[4.14] P. G. CIARLET, M. H. SCHULTZ AND R. S. VARGA, *Numerical methods of high-order accuracy for nonlinear boundary value problems. V. Monotone operator theory*, Numer. Math., 13 (1969), pp. 51–77.

CHAPTER 5

Fourier Analysis

5.1. General remarks and notation. For a bounded region Ω of R^n, our previous uses of piecewise-polynomial functions in approximating solutions of nonlinear elliptic boundary value problems were restricted rather severely by the *exact* treatment of boundary conditions. If we now consider analogous problems defined over *all* R^n, this difficulty of satisfying boundary conditions is eliminated, and more importantly, the powerful tool of Fourier transforms can be employed. In this portion of the notes, we describe in part the extremely elegant and penetrating analysis of Strang and Fix [5.1], [5.2] and [5.3], which is related to the methods of di Guglielmo [5.4], Aubin [5.5] and Babuška [5.6].

As usual, let $C_0^\infty(R^n)$ denote the set of all $u(x) = u(x_1, \cdots, x_n)$ infinitely differentiable in R^n with compact support, i.e., $u(x)$ vanishes identically for all $|x| \equiv (\sum_{i=1}^n x_i^2)^{1/2}$ sufficiently large. Then for any nonnegative integer s,

$$\|u\|_{W_2^s(R^n)}^2 \equiv \sum_{|\alpha| \leq s} \|D^\alpha u\|_{L_2(R^n)}^2,$$

(5.1.1)

$$\|u\|_{W_\infty^s(R^n)} \equiv \sum_{|\alpha| \leq s} \|D^\alpha u\|_{L_\infty(R^n)}$$

define norms on $C_0^\infty(R^n)$, and $W_2^s(R^n)$ and $W_\infty^s(R^n)$ denote the completions of $C_0^\infty(R^n)$ with respect to these norms. For convenience, we write W_p^s for $W_p^s(R^n)$.

For $u \in C_0^\infty(R^n)$, let $\hat{u}$ denote its Fourier transform:

$$(5.1.2) \qquad \hat{u}(\xi) = \int_{R^n} u(x) e^{-ix\xi}\, dx,$$

where $x = (x_1, x_2, \cdots, x_n)$, $\xi = (\xi_1, \xi_2, \cdots, \xi_n)$, and $x\xi \equiv x_1\xi_1 + \cdots + x_n\xi_n$. It should be noted that when $u \in (W_2^p)_0$, its Fourier transform $\hat{u}$, by the Paley–Wiener theorem, is an *entire function of exponential type* (cf. Hörmander [5.7, p. 21]). This will be useful in the proof of Theorem 5.1. Next, using Parseval's formula, one can define a norm equivalent to $\|\cdot\|_{W_2^s}$ by means of

$$(5.1.3) \qquad \|u\|_{W_2^s}^2 = \int_{R^n} |\hat{u}(\xi)|^2 (1 + |\xi|^2)^s\, d\xi.$$

For other standard notation, let Z^n denote all n-tuples $(j_1, j_2, \cdots, j_n)$ of integers, with Z^n_+ denoting all n-tuples of nonnegative integers. For $\alpha, \beta \in Z^n_+$, define

$$|\alpha| = \alpha_1 + \alpha_2 + \cdots + \alpha_n, \quad x^\alpha = x_1^{\alpha_1} x_2^{\alpha_2} \cdots x_n^{\alpha_n},$$

$$\alpha! = \alpha_1! \cdots \alpha_n!, \quad \binom{\alpha}{\beta} = \frac{\alpha!}{\beta!(\alpha - \beta)!} \quad \text{for} \quad \alpha \geqq \beta,$$

where $\quad \alpha \geqq \beta \quad$ if and only if $\quad \alpha_v \geqq \beta_v, \quad 1 \leqq v \leqq n.$

The starting point of the investigation of Strang and Fix is a *fixed* function $\phi(x) \in (W_2^p)_0$, i.e., ϕ is an element of W_2^p with compact support. For a parameter $h > 0$ and any $j \in Z^n$, we can define from ϕ other functions ϕ_j^h in $(W_2^p)_0$ by means of

$$(5.1.4) \qquad \phi_j^h(x) = h^{-n/2} \phi\left(\frac{x}{h} - j\right).$$

For example, if

$$\sigma(x) = \begin{cases} 1 - x, & 0 \leqq x \leqq 1, \\ x + 1, & -1 \leqq x \leqq 0, \\ 0, & \text{otherwise}, \end{cases}$$

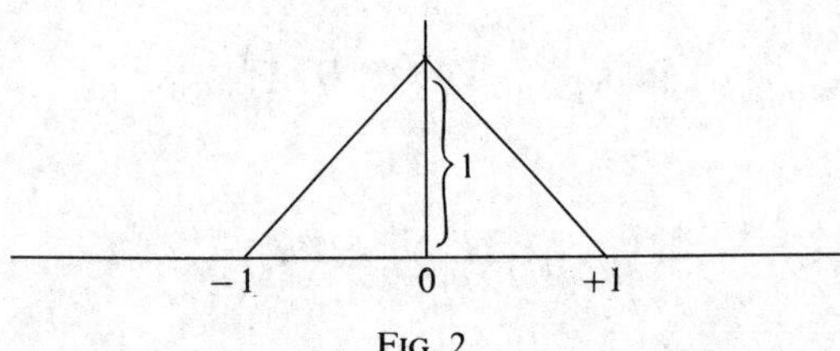

Fig. 2

then $\sigma \in (W_2^1(R))_0$, and the associated $\sigma_j^h(x)$ are the so-called "chapeau functions" (see Fig. 2). With these $\phi_j^h(x)$, we consider approximations in W_2^p by means of weighted sums of the ϕ_j^h:

$$(5.1.5) \qquad \sum_{j \in Z^n} w_j^h \phi_j^h(x).$$

The next result, an equivalence theorem, is one of the fundamental results of Strang and Fix [5.2].

THEOREM 5.1. *Let* $\phi \in (W_2^p)_0$. *Then the following conditions are equivalent*:
(i) $\hat{\phi}(0) \neq 0$, *but* $\hat{\phi}$ *has zeros of order at least* $p + 1$ *at the other points of* $2\pi Z^n$, *i.e.*,

$$(5.1.6) \qquad D^\alpha \hat{\phi}(2\pi j) = 0 \quad \text{if} \quad 0 \neq j \in Z^n, \quad |\alpha| \leqq p;$$

(ii) *for any* $|\alpha| \leqq p$, $\sum_{j \in Z^n} j^\alpha \phi(t - j)$ *is a polynomial in* $t_1, \cdots, t_n$ *with leading coefficient* Ct^α, $C \neq 0$;

(iii) *for any $u \in W_2^{p+1}$, there exist weights w_j^h such that as $h \to 0$,*

$$(5.1.7) \qquad \left\| u - \sum_{j \in Z^n} w_j^h \phi_j^h \right\|_{W_2^s} \leqq c_s h^{p+1-s} \|u\|_{W_2^{p+1}}, \qquad 0 \leqq s \leqq p,$$

with

$$(5.1.8) \qquad \sum_{j \in Z^n} |w_j^h|^2 \leqq K \|u\|_{W_2^0}^2,$$

where the constants c_s and K are independent of u.

The proof of Theorem 5.1 is too lengthy to reproduce here. We do remark that showing the equivalence of (i) and (ii) of Theorem 5.1 depends on the Poisson formula (cf. [5.7]) and is relatively easy. The equivalence of (i) with (iii) is much harder; one needs here the Paley–Wiener theorem and certain sharp estimates from the theory of entire functions (cf. [5.7, p. 21]).

It is important also to mention that if $u \in W_2^{p+1}$ has compact support, then the approximation $\sum_{j \in Z^n} w_j^h \phi_j^h$ in (5.1.7) *also* has compact support, and is supported in fact in

$$\left\{ x \in R^n : \text{dist}\,(x, \text{supp}\,u) \leqq \left(\frac{p+1}{2}\right) h \right\}.$$

This in particular means that Theorem 5.1, with its approximation theoretic result of (5.1.7), can be applied to problems *with boundaries*, such as (4.3.1), provided that homogeneous data is given on the boundary. Next, it is also important to mention that the exponent of h in (5.1.7) is *sharp*; moreover, the error bound of (5.1.7) is valid for *nonintegral* values of s. This latter remark is a consequence of the fact that the expression in (5.1.3) is a norm for nonintegral values of s.

Some interesting observations can be made from Theorem 5.1. In one dimension ($n = 1$), the simplest function which satisfies (i) of Theorem 5.1 is

$$(5.1.9) \qquad \hat{\sigma}_p(\xi) = \left\{ \frac{\sin\,(\xi/2)}{\xi/2} \right\}^{p+1},$$

which, as it turns out, is the Fourier transform of the Schoenberg *B-spline* $\sigma_p(x)$ (cf. Schoenberg [5.8]). For $p = 1$, $\sigma_1(x)$ is the chapeau function previously considered; for $p = 3$, $\sigma_3(x)$ is the *cubic spline*, given by

$$(5.1.10) \quad \sigma_3(x) = \frac{1}{6} \begin{cases} 0, & x \geqq 2, \\ (2 - x)^2, & 1 \leqq x \leqq 2, \\ 1 + 3(1 - x) + 3(1 - x)^2 - 3(1 - x)^3, & 0 \leqq x \leqq 1, \\ \sigma_3(-x), & x \leqq 0. \end{cases}$$

As is well known, $\sigma_3(x)$, the cubic B-spline, turns out to be very useful in practical computations (cf. Herbold [5.9]) for generating a basis for spline calculations.

A generalized form of Theorem 5.1 comes about from considerations of Hermite splines in one dimension. As we have seen in one dimension ($n = 1$), Hermite splines have several basic functions per knot. This suggests considering N fixed functions $\phi_1, \cdots, \phi_N \in (W_2^p)_0$ and defining for each parameter $h > 0$, the functions

$$(5.1.11) \qquad \phi_{i,j}^h(x) = h^{-n/2}\phi_i\left(\frac{x}{h} - j\right), \qquad 1 \leq i \leq N, \quad j \in Z^n.$$

Approximations in W_2^p are now to be obtained by means of sums of the form

$$(5.1.12) \qquad \sum_{\substack{1 \leq i \leq N \\ j \in Z^n}} w_{i,j}^h \phi_{i,j}^h(x).$$

A generalization of Theorem 5.1 given by Strang and Fix [5.2] is as follows.

THEOREM 5.2. *Let* $\phi_1, \phi_2, \cdots, \phi_N \in (W_2^q)_0$ *and let* $p \geq q \geq 0$. *Then the following are equivalent*:

(i) *There are linear combinations* ψ_α *of the* ϕ_i *which satisfy*

$$(5.1.13) \qquad \hat{\psi}_0(0) = 1, \quad \hat{\psi}_0(2\pi j) = 0 \quad \text{for all} \quad 0 \neq j \in Z^n,$$

and

$$(5.1.14) \qquad \sum_{\beta \leq \alpha} \frac{D^\beta \hat{\psi}_{\alpha-\beta}(2\pi j)}{\beta! i^{|\beta|}} = 0 \quad \text{for all} \quad j \in Z^n, \qquad 1 \leq |\alpha| \leq p;$$

(ii) *there are linear combinations* ψ_α *of the* ϕ_i *which satisfy*

$$(5.1.15) \qquad \frac{t^\alpha}{\alpha!} = \sum_{\beta \leq \alpha} \sum_j \frac{j^\beta \psi_{\alpha-\beta}(t - j)}{\beta!}, \qquad |\alpha| \leq p;$$

(iii) *for each* $u \in W_2^{p+1}$, *there are weights* $w_{i,j}^h$ *such that as* $h \to 0$,

$$(5.1.16) \quad \|u - \sum_{i,j} w_{i,j}^h \phi_{i,j}^h\|_{W_2^s} \leq c_s h^{p+1-s}\|u\|_{W_2^{p+1}} \quad \text{for} \quad s = 0, 1, \cdots, q,$$

with

$$(5.1.17) \qquad \sum_{i,j} |w_{i,j}^h|^2 \leq K\|u\|_{W_2^0}^2,$$

the constants c_s *and* K *are independent of* u.

Even in the uniform norm, Strang and Fix [5.2] have obtained results which generalize known results for spline approximation with respect to uniform partitions Δ_u.

THEOREM 5.3. *Let* $\phi_1, \cdots, \phi_N$ *satisfy the conditions of Theorem 5.2, and assume moreover that* $\phi_1, \cdots, \phi_N \in (W_\infty^q)_0$, *and let* $0 \leq q \leq p$. *Define the quasi-interpolate of* $u \in W_\infty^{p+1}$ *by* (cf. Theorem 5.2, (i))

$$(5.1.18) \qquad \tilde{u}^h(x) = \sum_{j \in Z^n} \sum_{|\alpha| \leq p} (D^\alpha u)(jh)h^{|\alpha|}\psi_\alpha\left(\frac{x}{h} - j\right).$$

Then

$$(5.1.19) \qquad \|u - \tilde{u}^h\|_{W_\infty^s} \leq c_s h^{p+1-s}\|u\|_{W_\infty^{p+1}} \quad \text{for} \quad s = 0, 1, \cdots, q.$$

The basic idea used in proving Theorem 5.3 is a finite Taylor expansion of $\tilde{u}^h(x)$, coupled with the polynomial generating property (5.1.15) of the ψ_α's. More precisely, writing $x/h = k + t$, where $t = (t_1, \cdots, t_n)$ with $0 \leq t_v < 1$, then with $l = j - k$, (5.1.18) gives

$$D^\beta \tilde{u}^h(x) = \sum_l \sum_{|\alpha| \leq p} D^\alpha u(kh + lh) h^{|\alpha| - |\beta|} D_t^\beta \psi_\alpha(t - l).$$

Expanding $D^\alpha u(kh + lh)$ in a finite Taylor series through order p gives

$$D^\beta \tilde{u}^h(x) = \sum_l \sum_{|\alpha + \gamma| \leq p} D^{\alpha + \gamma} u(kh) \frac{(lh)^\gamma}{\gamma!} h^{|\alpha| - |\beta|} D_t^\beta \psi_\alpha(t - l) + E_1(x).$$

Using (5.1.15), this can be written for $|\beta| = s \leq p$, after some manipulation, as

$$D^\beta \tilde{u}^h(x) = D^\beta u(x) + E_2(x),$$

where $E_2(x)$ is bounded by the right-hand side of (5.1.19).

It is interesting to mention other analogues of (5.1.19). Given any open set $\Omega \subset R^n$, let $\Omega' \subset \Omega$ be such that

$$\text{dist}(\Omega', \partial\Omega) \geq (S + \sqrt{n})h,$$

where S is the radius of the smallest sphere centered at the origin outside which all ϕ_i vanish. Then, a quasi-interpolate $\tilde{u}^h$ of u can be found such that

$$(5.1.20) \qquad \|u - \tilde{u}^h\|_{W_\infty^s(\Omega')} \leq c_s h^{p+1-s} \|u\|_{W_\infty^{p+1}(\Omega)} \quad \text{for} \quad s = 0, 1, \cdots, q.$$

For details, see Strang and Fix [5.2], Descloux [5.10], and de Boor and Fix [5.12].

To give a concrete form of the quasi-interpolation of (5.1.18) of Theorem 5.3, we have for $n = 1, N = 1, q = 3 = p$, that

$$(5.1.21) \qquad \tilde{u}^h(x) = \sum_{j \in Z} \left\{ u(jh) - \frac{h^2}{6} D^2 u(jh) \right\} \sigma_3 \left(\frac{x}{h} - j \right),$$

where $\sigma_3(x)$ is defined in (5.1.10). Because σ_3 has its support on $[-2, +2]$ from (5.1.10), and $\sigma_3(0) = \frac{4}{6}$, $\sigma_3(\pm 1) = \frac{1}{6}$, then (5.1.21) reduces for $x = 0$ to

$$\tilde{u}^h(0) = \frac{1}{6} \left\{ (4u(0) + u(h) + u(-h)) - \frac{h^2}{6} [4D^2 u(0) + D^2 u(h) + D^2 u(-h)] \right\},$$

and simple manipulations involving finite Taylor series expansion show that

$$(5.1.22) \qquad \tilde{u}^h(0) = u(0) + O(h^4)$$

for $u \in W_\infty^4(R)$. Note that $\tilde{u}^h$ does not in general interpolate $u(x)$ at the points $jh, j \in Z$; hence the name *quasi-interpolate*.

5.2. Applications. Because of the powerful nature of the results of § 5.1, there are many particular places where these error bounds can be applied. The first such application would most naturally be to approximations of solutions of elliptic problems by means of Galerkin methods.

To begin, we consider the linear elliptic problem over all R^n:

$$(5.2.1) \qquad\qquad \mathscr{L}u(x) = f(x), \qquad\qquad x \in R^n,$$

where

$$(5.2.2) \qquad\qquad \mathscr{L}u(x) \equiv \sum_{|\alpha|,|\beta| \le m} (-1)^\beta D^\beta \{ q_{\alpha\beta}(x) D^\alpha u(x) \},$$

where the coefficients $q_{\alpha\beta}(x)$ are all bounded in R^n, and where the associated bilinear form $B(u, v)$, defined by

$$(5.2.3) \qquad\qquad B(u, v) = \sum_{|\alpha|,|\beta| \le m} \int_{R^n} q_{\alpha\beta} D^\alpha u D^\beta v \, dx, \qquad\qquad u, v \in W_2^m,$$

is $W_2^m(R^n)$-*elliptic*, i.e., there is a constant $\rho > 0$ such that

$$(5.2.4) \qquad\qquad B(u, u) \ge \rho \|u\|_{W_2^m}^2 \quad \text{for all} \quad u \in W_2^m.$$

Then, u is a generalized solution of (5.2.1) if

$$(5.2.5) \qquad\qquad B(u, v) = \int_{R^n} fv \, dx \quad \text{for all} \quad v \in W_2^m.$$

Similarly, if S^h is a finite-dimensional subspace of W_2^m, then u^h is the Galerkin approximation to the solution u of (5.2.1) if

$$(5.2.6) \qquad\qquad B(u^h, w^h) = \int_{R^n} fw^h \, dx \quad \text{for all} \quad w^h \in S^h.$$

From (5.2.5) and (5.2.6), it follows that

$$(5.2.7) \qquad\qquad B(u^h - u, w^h) = 0 \quad \text{for all} \quad w^h \in S^h.$$

Hence, from (5.2.4), (5.2.7), and the boundedness of the coefficients in $B(u, v)$,

$$\rho \|u^h - u\|_{W_2^m}^2 \le B(u^h - u, u^h - u) = B(u^h - u, w^h - u)$$

$$\le K \|u^h - u\|_{W_2^m} \|w^h - u\|_{W_2^m} \quad \text{for all } w^h \in S^h.$$

Consequently,

$$(5.2.8) \qquad\qquad \|u^h - u\|_{W_2^m} \le K \inf_{w^h \in S^h} \|u - w^h\|_{W_2^m}.$$

We point out that the above inequality could have been deduced *directly* as a special case of (4.2.8) of the more general Theorem 4.5 on monotone nonlinear operators.

If S^h is composed of elements of the form (5.1.12), where the ϕ_i's satisfy condition (i) of Theorem 5.2 with $q \ge m$, then we can use the error bound of (5.1.16), i.e., if $u \in W_2^{p+1}$, $p + 1 \ge m$, then

$$(5.2.9) \qquad\qquad \|u^h - u\|_{W_2^m} \le K h^{p+1-m} \|u\|_{W_2^{p+1}}.$$

To obtain improved error bounds in the norm $\|u^h - u\|_{W_2^s}$ for $0 \leq s \leq m$ is less obvious. The following result is due to Strang and Fix [5.2].

THEOREM 5.4. *Let the generalized solution u of (5.2.5) be in $W_2^{p+1}, p + 1 \geq m$. If the subspace S^h of W_2^m satisfies hypothesis* (i) *of Theorem 5.2, then the Galerkin approximation u^h in S^h (cf. (5.2.6)) satisfies*

$$(5.2.10) \qquad \|u - u^h\|_{W_2^s} \leq Kh^\sigma \|u\|_{W_2^{p+1}}, \qquad\qquad 0 \leq s \leq m,$$

where

$$(5.2.11) \qquad \sigma \equiv \min\{p + 1 - s; 2(p + 1 - m)\}.$$

Similar results in the norm $\| \cdot \|_{W_\infty^s}$ are also in [5.2]. It is also shown in [5.2] that the exponent of h in (5.2.10), given by (5.2.11), is best possible.

We remark that the error bounds of (5.2.10) and (5.2.11) could also have been derived from a technique due to Nitsche [5.11], and hence, the exponent of h in (5.2.10), as given by (5.2.11), remains sharp for problems with boundaries as well.

REFERENCES

[5.1] G. FIX AND G. STRANG, *Fourier analysis of the finite element method in Ritz-Galerkin theory*, Studies in Appl. Math., 48 (1969), pp. 265–273.

[5.2] G. STRANG AND G. FIX, *An Analysis of the Finite Element Method*, Prentice-Hall, Englewood Cliffs, N.J., to appear.

[5.3] G. STRANG, *The finite element method and approximation theory*, Numerical Solution of Partial Differential Equations, II, B. E. Hubbard, ed., Academic Press, New York, 1971, pp. 547–583.

[5.4] F. DIGULIELMO, *Construction d'approximations des espaces de Sobolev sur des réseaux en simplexes*, Calcolo, 6 (1969), pp. 279–331.

[5.5] J.-P. AUBIN, *Evaluation des erreurs de troncature des approximations des espaces de Sobolev*, J. Math. Anal. Appl., 21 (1968), pp. 356–368.

[5.6] I. BABUŠKA, *Approximation by hill functions*, Tech. Note BN-648, Institute for Fluid Dynamics and Applied Mathematics, University of Maryland, 1970.

[5.7] L. HÖRMANDER, *Linear Partial Differential Operators*, Academic Press, New York, 1963.

[5.8] I. J. SCHOENBERG, *Contributions to the problem of approximation of equidistant data by analytic functions, Parts A and B*, Quart. Appl. Math., 4 (1946), pp. 45–99 and pp. 112–141.

[5.9] R. J. HERBOLD, *Consistent quadrature schemes for the numerical solution of boundary value problems by variational techniques*, Thesis, Case Western Reserve University, 1968.

[5.10] J. DESCLOUX, *Some properties of approximations by finite elements*, Report, Ecole Polytechnique Federale Lausanne, 1969.

[5.11] J. NITSCHE, *Ein Kriterium für die Quasi-Optimalität des Ritzschen Verfahrens*, Numer. Math., 11 (1968), pp. 346–348.

[5.12] C. DE BOOR AND G. FIX, *Spline approximation by quasi-interpolants*, J. Approx. Theory, to appear.

CHAPTER 6

Least Squares Methods

6.1. General remarks and notation. If Ω is a bounded region in R^n with boundary $\partial\Omega$, one can seek to approximate the solution of the elliptic differential equation:

$$(6.1.1) \qquad Au \equiv - \sum_{i,j=1}^{n} \frac{\partial}{\partial x_i}\left(a_{i,j}(x)\frac{\partial u}{\partial x_j}\right) = f(x), \qquad x \in \Omega,$$

subject to boundary conditions of

$$(6.1.2) \qquad u = g, \quad x \in \partial\Omega,$$

in a number of ways. Thus, for the case of homogeneous boundary conditions, i.e.,

$$(6.1.2') \qquad u = 0, \quad x \in \partial\Omega,$$

the Galerkin method, as we have seen, can be described as follows. Defining the bilinear form $B(u, v)$ by

$$B(u, v) \equiv \sum_{i,j=1}^{n} \int_\Omega a_{i,j}(x)\frac{\partial u}{\partial x_j}\frac{\partial v}{\partial x_i}\,dx \quad \text{for all} \quad u, v \in \mathring{W}_2^1(\Omega),$$

then the unique (with suitable regularity assumptions) generalized solution u of (6.1.1)–(6.1.2') satisfies

$$(6.1.3) \qquad B(u, v) = (f, v)_0 \equiv \int_\Omega fv\,dx \quad \text{for all} \quad v \in \mathring{W}_2^1(\Omega).$$

If S^M is any finite-dimensional subspace of $\mathring{W}_2^1(\Omega)$, then the Galerkin approximation $\hat{w}$ in S^M to u analogously is defined by

$$(6.1.4) \qquad B(\hat{w}, v) = (f, v)_0 \quad \text{for all} \quad v \in S^M.$$

Another approach is the *least squares method*. Here, one defines the least squares approximation $\tilde{w}$ in S^M of the solution u of (6.1.1)–(6.1.2) as that $w \in S^M$ which minimizes

$$(6.1.5) \qquad \|Aw - f\|_{L_2(\Omega)}^2 + K_M\|w - g\|_{L_2(\partial\Omega)}^2,$$

where

$$\|w - g\|_{L_2(\partial\Omega)}^2 \equiv \int_{\partial\Omega} (w - g)^2\,d\sigma.$$

The constant K_M in general is *large* and will depend upon the choice of the subspace S^M; for example, K_M is chosen in (6.2.6) to be h^{-3}.

One very important feature of this method is that it is *not necessary* for the elements of the subspace S^M to satisfy the boundary conditions of (6.1.2). One possible drawback, of computation importance, is that the associated matrix, which is used to determine $\tilde{w}$ in (6.1.5), has a condition number roughly the square of that for the corresponding matrix of the Galerkin method. Another drawback, of lesser importance, is the necessity in the least squares method of working with subspaces S^M having smoother elements than that required by the Galerkin method; the Galerkin method requires from (6.1.4) that S^M be a subspace of $\mathring{W}_2^1(\Omega)$, whereas the least squares method requires from (6.1.5) that $Aw \in L_2(\Omega)$ for each $w \in S^M$, i.e., $S^M \subset W_2^2(\Omega)$.

In this portion of the notes, we shall describe in part the recent work of Bramble and Schatz [6.1]–[6.3] on such least squares methods. In so doing, we need some extra notation, which is described below.

Let Ω be a bounded region in R^n with a boundary $\partial\Omega$ which is sufficiently smooth; for convenience, assume $\partial\Omega$ is of class C^∞. As before, if p is any nonnegative integer, the Hilbert space $W_2^p(\Omega)$ is the completion of $C^\infty(\overline{\Omega})$ in the norm

$$\|u\|_{W_2^p(\Omega)}^2 \equiv \sum_{|\alpha| \leqq p} \|D^\alpha u\|_{L_2(\Omega)}^2,$$

and if p is any positive real number which is not an integer, say $j < p < j + 1$, then the Hilbert space $W_2^p(\Omega)$ is defined as the interpolation (as described in § 1.2) between the spaces $W_2^j(\Omega)$ and $W_2^{j+1}(\Omega)$, i.e., if $p = j + \theta, 0 < \theta < 1$, then (cf. (1.2.13)), with equivalence of norms,

$$(6.1.6) \qquad W_2^p(\Omega) \equiv (W_2^j(\Omega), W_2^{j+1}(\Omega))_{\theta,2}.$$

Next, we need from the *theory of traces* the notion of the Hilbert spaces $W_2^p(\partial\Omega)$, defined on the boundary $\partial\Omega$ of Ω. For the particular case $p = 0$, the elements of $W_2^0(\partial\Omega)$ are just those functions v defined on $\partial\Omega$ such that

$$(6.1.7) \qquad \int_{\partial\Omega} |v|^2 \, d\sigma < \infty,$$

where σ is the measure on $\partial\Omega$ induced by Lebesgue measure in R^{n-1}, and thus $W_2^0(\partial\Omega) = L_2(\partial\Omega)$. For the case when p is an arbitrary real number, the description of $W_2^p(\partial\Omega)$ is somewhat more complicated. To simplify our discussion here, we state that a norm $|\cdot|_p$ and an inner product $\langle \cdot, \cdot \rangle_p$ can be defined on $C^\infty(\partial\Omega)$ (cf. Nečas [6.4], Lions and Magenes [6.5] and Bramble and Schatz [6.1]) such that $W_2^p(\partial\Omega)$ is then defined as the completion of $C^\infty(\partial\Omega)$ in this norm $|\cdot|_p$.

Next, let $H^{(l,s)}$ be the Hilbert space defined as the Cartesian product $W_2^l(\Omega) \times W_2^s(\partial\Omega)$, with inner product defined by

$$(6.1.8) \qquad (F_1, F_2)_{(l,s)} \equiv (f_1, f_2)_l + \langle g_1, g_2 \rangle_s$$

and norm defined by

$$(6.1.9) \qquad \|F_1\|_{(l,s)}^2 \equiv \|f_1\|_l^2 + |g_1|_s^2,$$

where $F_1 = \{f_1, g_1\}$, $F_2 = \{f_2, g_2\}$, and $(\cdot, \cdot)_l$ and $\|\cdot\|_l$ denote the inner product and norm in $W_2^l(\Omega)$. In place of the differential operator of (6.1.1), consider now the more general second order differential operator

$$(6.1.10) \qquad Au = -\sum_{i,j=1}^{n} D_i a_{i,j}(x) D_j u + \sum_{i=1}^{n} a_i(x) D_i u + a(x)u, \qquad D_i \equiv \frac{\partial}{\partial x_i},$$

where all coefficients are of class $C^\infty(R^n)$, and assume that A is *uniformly elliptic* in $\bar{\Omega}$, i.e., there exists a constant $C > 0$ such that

$$(6.1.11) \qquad C^{-1}|\xi|^2 \leqq \left| \sum_{i,j=1}^{n} a_{i,j}(x)\xi_i\xi_j \right| \leqq C|\xi|^2$$

for all $x \in \bar{\Omega}$ and all $\xi \in R^n$, where $|\xi|^2 = \xi_1^2 + \cdots + \xi_n^2$, and assume that

$$(6.1.12) \qquad \begin{array}{l} \text{the only solution in } C^\infty(\bar{\Omega}) \text{ of } Au = 0 \text{ in } \Omega, \\ u = 0 \text{ on } \partial\Omega, \text{ is the zero solution.} \end{array}$$

With the assumptions of (6.1.11)–(6.1.12), Lions and Magenes [6.5] have shown that, for any real number $p \geqq 2$, the norms

$$\|\cdot\|_{W_2^p(\Omega)} \quad \text{and} \quad \left\{ \|A\cdot\|^2_{W_2^{p-2}(\Omega)} + |\cdot|^2_{W_2^{p-1/2}(\partial\Omega)} \right\}^{1/2}$$

are *equivalent* on $W_2^p(\Omega)$, i.e., there exist positive constants C_1 and C_2 (independent of u) such that

$$(6.1.13) \qquad \|u\|_{W_2^p(\Omega)} \leqq C_1 \left\{ \|Au\|^2_{W_2^{p-2}(\Omega)} + |u|^2_{W_2^{p-1/2}(\partial\Omega)} \right\}^{1/2} \leqq C_2 \|u\|_{W_2^p(\Omega)}$$

for all $u \in W_2^p(\Omega)$. In other words, the mapping $u \to \{Au, u\}$ is a homeomorphism of $W_2^p(\Omega)$ onto $H^{(p-2, p-1/2)}$ for $p \geqq 2$. We also remark that the first inequality of (6.1.13) is valid for all real p (cf. [6.1]).

Next, consider any $\{f, g\} \in H^{(p-2, p-1/2)}$ for any real p. By definition of the completion of the spaces considered, there exists a sequence $\{\{f_n, g_n\}\}_{n=1}^{\infty}$ with $\{f_n, g_n\} \in C^\infty(\bar{\Omega}) \times C^\infty(\partial\Omega)$ for each $n \geqq 1$ such that $\{f_n, g_n\}$ converges to $\{f, g\}$ in the norm $\|\cdot\|_{(p-2, p-1/2)}$ as $n \to \infty$. Because of the smoothness assumptions on the coefficients of A in (6.1.10) and the assumptions of (6.1.11)–(6.1.12), for each $n \geqq 1$, there necessarily exists a unique $u_n \in C^\infty(\bar{\Omega})$ such that

$$(6.1.14) \qquad \begin{array}{l} Au_n = f_n \quad \text{in} \quad \Omega, \\[4pt] u_n = g_n \quad \text{on} \quad \partial\Omega. \end{array}$$

Hence, from the first inequality of (6.1.13) and (6.1.14), for any m and n, we have

$$\begin{aligned} \|u_m - u_n\|_{W_2^p(\Omega)} &\leqq C_1 \left\{ \|A(u_m - u_n)\|^2_{W_2^{p-2}(\Omega)} + |u_m - u_n|^2_{W_2^{p-1/2}(\partial\Omega)} \right\}^{1/2} \\ &= C_1 \left\{ \|f_m - f_n\|^2_{W_2^{p-2}(\Omega)} + |g_m - g_n|^2_{W_2^{p-1/2}(\partial\Omega)} \right\}^{1/2} \\ &\equiv C_1 \|\{f_m, g_m\} - \{f_n, g_n\}\|_{(p-2, p-1/2)}. \end{aligned}$$

Thus, since $\{f_n, g_n\} \to \{f, g\}$, then $\{u_n\}_{n=1}^{\infty}$ is a Cauchy sequence in $W_2^p(\Omega)$, and we define the unique limit, u, of $\{u_n\}_{n=1}^{\infty}$ to be the generalized solution in $W_2^p(\Omega)$ of

$$Au = f \quad \text{in} \quad \Omega,$$

(6.1.15)

$$u = g \quad \text{on} \quad \partial\Omega.$$

This will be useful in the next section.

6.2. Approximation theoretic results. Let $S_{k,r}^h(R^n)$ be any finite-dimensional subspace of $W_2^k(R^n)$, where k and r are nonnegative integers with $k < r$, and h is a parameter with $0 < h \leq 1$. We assume that $S_{k,r}^h(R^n)$ has the approximation property such that for any $v \in W_2^r(R^n)$, there is a $w \in S_{k,r}^h(R^n)$ such that

$$(6.2.1) \qquad \|v - w\|_{W_2^k(R^n)} \leq Ch^{r-k}\|v\|_{W_2^r(R^n)},$$

where C is independent of h and v. It is clear from the equivalence theorems of Strang and Fix (cf. Theorems 5.1 and 5.2) and the approximation results of Bramble and Hilbert (cf. Theorems 3.3 and 3.5) that such subspaces $S_{k,r}^h(R^n)$ are in fact easily generated.

Next, let $S_{k,r}^h(\Omega)$ denote the restriction of $S_{k,r}^h(R^n)$ to functions defined on $\bar{\Omega}$. By Calderón's extension theorem (cf. Agmon [6.6, p. 171]), there exists a bounded linear transformation $\mathscr{E}: W_2^r(\Omega) \to W_2^r(R^n)$ with $\mathscr{E}v \equiv v$ on Ω, i.e., for some constant M,

$$(6.2.2) \qquad \|\mathscr{E}v\|_{W_2^r(R^n)} \leq M\|v\|_{W_2^r(\Omega)} \quad \text{for all} \quad v \in W_2^r(\Omega).$$

Thus, for any $v \in W_2^r(\Omega)$, we have $\mathscr{E}v \in W_2^r(R^n)$, and applying (6.2.1), there is a $w \in S_{k,r}^h(R^n)$ such that

$$\|\mathscr{E}v - w\|_{W_2^k(R^n)} \leq Ch^{r-k}\|\mathscr{E}v\|_{W_2^r(R^n)} \leq CMh^{r-k}\|v\|_{W_2^r(\Omega)},$$

the last inequality following from (6.2.2). But since, by definition, $\|v\|_{W_2^k(R^n)} \geq \|v\|_{W_2^k(\Omega)}$ for any $v \in W_2^k(R^n)$, and as $\mathscr{E}v \equiv v$ on Ω, the above inequality becomes

$$\|v - w\|_{W_2^k(\Omega)} \leq C'h^{r-k}\|v\|_{W_2^r(\Omega)}$$

for any $v \in W_2^r(\Omega)$. Equivalently, for any $v \in W_2^r(\Omega)$,

$$(6.2.3) \qquad \inf_{w \in S_{k,r}^h(\Omega)} \|v - w\|_{W_2^k(\Omega)} \leq C'h^{r-k}\|v\|_{W_2^r(\Omega)}.$$

Now, with $2 = k < r$, choose v in (6.2.3) to be the generalized solution, u, of (6.1.15). The right side of (6.2.3) can be bounded above from the first inequality of (6.1.13), i.e., using (6.1.15),

$$\|u\|_{W_2^r(\Omega)} \leq C_1\{\|Au\|_{W_2^{r-2}(\Omega)}^2 + |u|_{W_2^{r-1/2}(\partial\Omega)}^2\}^{1/2}$$

$$= C_1\{\|f\|_{W_2^{r-2}(\Omega)}^2 + |g|_{W_2^{r-1/2}(\partial\Omega)}^2\}^{1/2}.$$

Hence, combining the above inequality with that of (6.2.3) gives for $2 = k < r$,

$$(6.2.4) \qquad \inf_{w \in S_{2,r}^h(\Omega)} \|u - w\|_{W_2^2(\Omega)} \leq C''h^{r-2}\{\|f\|_{W_2^{r-2}(\Omega)}^2 + |g|_{W_2^{r-1/2}(\partial\Omega)}^2\}^{1/2},$$

which, from (6.1.15) and the equivalence relations in (6.1.13) for $p = 2$, we can write also as

$$(6.2.4')\quad
\begin{aligned}
\inf_{w \in S_{2,r}^h(\Omega)} & \left\{ \|f - Aw\|_{L_2(\Omega)}^2 + |g - w|_{W_2^{3/2}(\partial\Omega)}^2 \right\}^{1/2} \\
& \leq C''' h^{r-2} \left\{ \|f\|_{W_2^{r-2}(\Omega)}^2 + |g|_{W_2^{r-1/2}(\partial\Omega)}^2 \right\}^{1/2}.
\end{aligned}$$

Note that the right-hand side of the above inequalities (6.2.4) and (6.2.4') depends solely on the data of the problem. Moreover, the minimization problem as given on the left-hand side of (6.2.4'), while not of the form originally considered in (6.1.5), does from (6.2.4) give an error bound in the $W_2^2(\Omega)$-norm for this minimization problem. The object now is to mimic the inequalities of (6.2.4)–(6.2.4') for the minimization problem of (6.1.5), thereby obtaining error estimates for this least squares method.

First, Bramble and Schatz [6.1] have established the following result which is the desired analogue of (6.2.4'). It uses, in an iterated manner, the fundamental result of Theorem 1.7 on interpolation spaces.

THEOREM 6.1. *Let* $S_{k,r}^h(R^n)$ *satisfy* (6.2.1) *with* $2 = k < r$, *and suppose that* $\{f, g\} \in H^{(\lambda, \lambda_0)}$, *where* $0 \leq \lambda \leq r - 2$, *and* $0 \leq \lambda_0 \leq r - \frac{1}{2}$. *Then there exists a constant* C *independent of* $\{f, g\}$ *and* h *such that*

$$(6.2.5)\quad
\begin{aligned}
\inf_{w \in S_{2,r}^h(\Omega)} & \left\{ \|f - Aw\|_{L_2(\Omega)}^2 + h^{-3}|g - w|_{L_2(\partial\Omega)}^2 \right\}^{1/2} \\
& \leq C \left\{ h^{2\lambda} \|f\|_{W_2^\lambda(\Omega)}^2 + h^{2\lambda_0 - 3}|g|_{W_2^{\lambda_0}(\partial\Omega)}^2 \right\}^{1/2}.
\end{aligned}$$

Next, let us consider the minimization problem of (6.1.5) with $K_M = h^{-3}$ and $S^M \equiv S_{2,r}^h$. It is easily seen that minimizing

$$(6.2.6)\qquad \|f - Aw\|_{L_2(\Omega)}^2 + h^{-3}|g - w|_{L_2(\partial\Omega)}^2$$

to find $\tilde{w}$ in $S_{2,r}^h(\Omega)$ is equivalent to the orthogonality condition

$$(6.2.7)\quad \int_\Omega (f - A\tilde{w})Av\, dx + h^{-3}\int_{\partial\Omega}(g - \tilde{w})v\, d\sigma = 0 \quad \text{for all} \quad v \in S_{2,r}^h(\Omega),$$

which we can write equivalently as

$$(6.2.8)\qquad (f - A\tilde{w}, Av)_0 + h^{-3}\langle g - \tilde{w}, v \rangle_0 = 0 \quad \text{for all} \quad v \in S_{2,r}^h(\Omega),$$

using our previous notation. Next, to obtain one of the error bounds of Bramble and Schatz for the least squares method applied to (6.1.15), we need the following inequality (cf. [6.1, Lemma 2.1]):

$$(6.2.9)\qquad \|v\|_{L_2(\Omega)} \leq C \sup_{y \in C^\infty(\Omega)} \left\{ \frac{(Av, Ay)_0 + h^{-3}\langle v, y \rangle_0}{\left(\|Ay\|_{W_2^2(\Omega)}^2 + h^{-6}|y|_{W_2^{1/2}(\partial\Omega)}^2\right)^{1/2}} \right\}$$

for any $v \in L_2(\Omega)$. Thus, if $\tilde{w} \in S^h_{2,r}(\Omega)$ is the least squares approximation to the solution u of (6.1.15), then from (6.2.9),

$$(6.2.10) \qquad \|u - \tilde{w}\|_{L_2(\Omega)} \leq C \sup_{y \in C^\infty(\overline{\Omega})} \left\{ \frac{(f - A\tilde{w}, Ay)_0 + h^{-3}\langle g - \tilde{w}, y \rangle_0}{(\|Ay\|^2_{W^2_2(\Omega)} + h^{-6}|y|^2_{W^{1/2}_2(\partial\Omega)})^{1/2}} \right\}.$$

However, from the orthogonality condition of (6.2.8), it is clear that the numerator of the right-hand side of (6.2.10) can be replaced by

$$(f - A\tilde{w}, Ay - Aw)_0 + h^{-3}\langle g - \tilde{w}, y - w \rangle_0 \quad \text{for all} \quad w \in S^h_{2,r}(\Omega).$$

Applying Schwarz's inequality to the above expression results in the upper bound

$$\left\{ \|f - A\tilde{w}\|^2_{L_2(\Omega)} + h^{-3}|g - \tilde{w}|^2_{L_2(\partial\Omega)} \right\}^{1/2} \cdot \left\{ \|Ay - Aw\|^2_{L_2(\Omega)} + h^{-3}|y - w|^2_{L_2(\partial\Omega)} \right\}^{1/2},$$

which of course is valid for all $w \in S^h_{2,r}(\Omega)$, and hence, the infimum can be taken over $S^h_{2,r}(\Omega)$. But then, we can apply (6.2.5) with $f = Ay$, $g = y$, $\lambda = 2$, and $\lambda_0 = \frac{1}{2}$. This gives us that

$$\inf_{w \in S^h_{2,r}(\Omega)} \left\{ \|Ay - Aw\|^2_{L_2(\Omega)} + h^{-3}|y - w|^2_{L_2(\partial\Omega)} \right\}^{1/2}$$

$$\leq Ch^2 \left\{ \|Ay\|^2_{W^2_2(\Omega)} + h^{-6}|y|^2_{W^{1/2}_2(\partial\Omega)} \right\}^{1/2}.$$

Thus, when these bounds are substituted into the numerator of the right-hand side of (6.2.10), we simply have

$$(6.2.11) \qquad \|u - \tilde{w}\|_{L_2(\Omega)} \leq Ch^2 \left\{ \|f - A\tilde{w}\|^2_{L_2(\Omega)} + h^{-3}|g - \tilde{w}|^2_{L_2(\partial\Omega)} \right\}^{1/2}.$$

But since the choice $\tilde{w}$ minimizes (6.2.6) over $S^h_{2,r}(\Omega)$, we can again apply (6.2.5) to the right-hand side of (6.2.11). This gives then the following important result of Bramble and Schatz [6.1].

THEOREM 6.2. *With the hypotheses of Theorem 6.1, let $\tilde{w} \in S^h_{2,r}(\Omega)$ be the unique element which minimizes*

$$(6.2.12) \qquad \|f - Aw\|^2_{L_2(\Omega)} + h^{-3}|g - w|^2_{L_2(\partial\Omega)}$$

over $S^h_{2,r}(\Omega)$. Then, for $r \geq 4$,

$$(6.2.13) \qquad \|u - \tilde{w}\|_{L_2(\Omega)} \leq Ch^2 \left\{ h^\lambda \|f\|_{W^2_2(\Omega)} + h^{\lambda_0 - 3/2}|g|_{W^{1/2}_{20}(\partial\Omega)} \right\}.$$

Again, to emphasize the important aspects of this least squares method, no boundary conditions need be satisfied by the approximate solution, the operator A need not be self-adjoint, and finally, only L_2-data is required of f and g, not point values. While we have described here only results for second order differential operators, this material has been generalized to $2m$th order operators by Bramble and Schatz [6.2].

Special cases of (6.2.13) are worthy of comment. If cubic splines are used, i.e., $r = 4$ in (6.2.1), and the data $\{f, g\}$ of (6.1.15) are such that $f \in W^2_2(\Omega)$ and $g \in W^{7/2}_2(\partial\Omega)$, then from (6.2.13), $\|u - \tilde{w}\|_{L_2(\Omega)} = O(h^4)$. On the other hand, if the $\{f, g\}$ are such that $f \in L_2(\Omega)$, $g \in L_2(\partial\Omega)$, then again from (6.2.13), $\|u - \tilde{w}\|_{L_2(\Omega)} = O(h^{1/2})$. Interior estimates of $u - \tilde{w}$ in different norms are also available (cf. [6.1]).

It is also necessary to mention that others have considered closely related methods. Babuška [6.7] uses a boundary perturbation in such a way that the trial functions need not satisfy boundary conditions, and similar penalty methods have been considered by Aubin [6.8]. These are important techniques, both theoretically as well as numerically, but space here does not permit a detailed description of such ideas. It should, however, be pointed out that the assumption that the boundary $\partial\Omega$ be smooth is a *critical* assumption for the analyses of Bramble and Schatz. It appears that the error estimates derived for the least squares approximations may break down for regions with corners, even if the solutions are smooth. Moreover, the least squares method may not yield good approximations to derivatives over all the region. For such reasons, the penalty function methods of Aubin and Babuška may prove to be in some cases more useful.

It would appear that the results mentioned in this section are of a purely theoretical nature only. However, numerical experiments with the least squares method are currently being conducted at Cornell University, and preliminary results are very encouraging, i.e., indications are that it is a numerically competitive method, as compared with the Galerkin method.

REFERENCES

[6.1] J. H. BRAMBLE AND A. H. SCHATZ, *Rayleigh–Ritz–Galerkin methods for Dirichlet's problem using subspaces without boundary conditions*, Comm. Pure Appl. Math., 23 (1970), pp. 653–676.

[6.2] ———, *Least squares method for 2mth order elliptic boundary value problems*, Math. of Comp., 25 (1971), pp. 1–32.

[6.3] ———, *On the numerical solution of elliptic boundary value problems by least squares approximation of the data*, Numerical Solution of Partial Differential Equations, II, B. E. Hubbard, ed., Academic Press, New York, 1971, pp. 107–131.

[6.4] J. NEČAS, *Les Méthodes Directes en Théorie des Équations Elliptiques*, Masson et Cie, Paris, 1967.

[6.5] J. L. LIONS AND E. MAGENES, *Problèmes aux Limites non Homogènes et Applications*, vol. 1, Dunod, Paris, 1968.

[6.6] S. AGMON, *Lectures on Elliptic Boundary Value Problems*, Van Nostrand, Princeton, New Jersey, 1965.

[6.7] I. BABUŠKA, *Numerical solution of boundary value problems by the perturbed variational principle*, Tech. Note BN-624, University of Maryland, 1969.

[6.8] J.-P. AUBIN, *Behavior of the error of the approximate solutions of boundary value problems for linear elliptic operators by Galerkin's and finite difference methods*, Annali della Scuola Normale di Pisa, 21 (1967), pp. 599–637.

CHAPTER 7

Eigenvalue Problems

7.1. The basic problem. The widespread need in many physical and engineering settings for *accurate* approximate eigenvalues has resulted in a long history of dedication by mathematicians and numerical analysts to the approximation of eigenvalues of general eigenvalue problems. Certainly, finite differences and Rayleigh–Ritz methods have been advocated for this purpose for many years (cf. Courant [7.1] and Kantorovich and Krylov [7.2]), but renewed interest in Rayleigh–Ritz methods using piecewise-polynomial function subspaces seems to have stemmed from the papers of Wendroff [7.3] and Birkhoff and de Boor [7.4]. In [7.3], $O(h^2)$ accuracy for the eigenvalues of Sturm–Liouville problems, using the subspaces $H^{(1)}(\Delta_u)$ of continuous piecewise-linear functions, was rigorously established. Then, in Birkhoff, de Boor, Swartz, and Wendroff [7.5], the subspaces $H^{(2)}(\Delta_u)$ and $Sp^{(2)}(\Delta_u)$ of piecewise *cubic* polynomials were used in a Rayleigh–Ritz setting for such (second order) Sturm–Liouville problems, with a resulting higher order $O(h^6)$ accuracy for the lowest eigenvalues. These results were then extended by Ciarlet, Schultz and Varga [7.6] to general one-dimensional eigenvalue problems, using L-splines in a Rayleigh–Ritz setting. Subsequent developments in higher dimensions have been considered by Schultz [7.7], Pierce and Varga [7.8], [7.9] and Birkhoff and Fix [7.10]. The last reference is particularly worthy of note, since it contains a large number of theoretical and numerical results.

To give the background for the eigenvalue problem, let Ω be a bounded region in R^n, $n \geq 1$, with boundary $\partial\Omega$. We seek to approximate the eigenvalues and eigenfunctions of the linear eigenvalue problem

$$(7.1.1) \qquad \mathfrak{N}u(x) = \lambda \mathfrak{M}u(x), \quad x \in \Omega \subset R^n,$$

subject to the homogeneous boundary conditions

$$(7.1.2) \qquad \mathscr{B}u(x) = 0, \quad x \in \partial\Omega,$$

where

$$(7.1.3) \qquad \mathfrak{N}u(x) \equiv \sum_{|\alpha| \leq m} (-1)^{|\alpha|} D^\alpha \{ p_\alpha(x) D^\alpha u(x) \}, \qquad x \in \Omega,$$

$$\mathfrak{M}u(x) \equiv \sum_{|\alpha| \leq r} (-1)^{|\alpha|} D^\alpha \{ q_\alpha(x) D^\alpha u(x) \}, \qquad x \in \Omega,$$

where $0 \leq r < m$, and where $\mathscr{B} = \{B_j\}_{j=1}^m$ consists of m linearly independent conditions of the form

$$(7.1.4) \qquad B_j u(x) \equiv \sum_{|\alpha| \leq 2m-1} \tau_{j,\alpha}(x) D^\alpha u(x) = 0, \qquad 1 \leq j \leq m, \quad x \in \partial\Omega.$$

In the case $n = 1$ with $\Omega = (a, b)$, we shall also allow the more general boundary conditions of the form

$$(7.1.5) \qquad B_j u(x) \equiv \sum_{k=1}^{2m} \{\tau_{j,k} D^{k-1} u(a) + \tau'_{j,k} D^{k-1} u(b)\} = 0, \qquad 1 \leq j \leq 2m.$$

Let $\mathfrak{D}$ be the linear space of all real-valued functions $u \in C^{2m}(\bar{\Omega})$ satisfying the boundary conditions of (7.1.4). We assume that

$$(7.1.6) \qquad \begin{aligned} (\mathfrak{N}u, v)_0 &= (u, \mathfrak{N}v)_0 \quad \text{for all} \quad u, v \in \mathfrak{D}, \\ (\mathfrak{M}u, v)_0 &= (u, \mathfrak{M}v)_0 \quad \text{for all} \quad u, v \in \mathfrak{D}, \end{aligned}$$

where $(u, v)_0 \equiv \int_\Omega uv \, dx$ is the usual L_2-inner product. In addition, we assume that there exist positive constants K_1 and K_2 for which

$$(7.1.7)$$

$$(\mathfrak{N}u, u)_0 \geq K_1 (\mathfrak{M}u, u)_0 \quad \text{and} \quad (\mathfrak{M}u, u)_0 \geq K_2 (u, u)_0 \quad \text{for all} \quad u \in \mathfrak{D}.$$

As in [7.6], we define the inner products

$$(7.1.8) \qquad (u, v)_D \equiv (\mathfrak{M}u, v)_0 \quad \text{for all} \quad u, v \in \mathfrak{D}$$

and

$$(7.1.9) \qquad (u, v)_N \equiv (\mathfrak{N}u, v)_0 \quad \text{for all} \quad u, v \in \mathfrak{D}.$$

As a consequence of (7.1.7), $\|u\|_D \equiv (u, u)_D^{1/2}$ and $\|u\|_N \equiv (u, u)_N^{1/2}$ are norms on $\mathfrak{D}$, and we denote the Hilbert space completions of $\mathfrak{D}$ with respect to $\|\cdot\|_D$ and $\|\cdot\|_N$, respectively as H_D and H_N. From (7.1.7) it follows that

$$(7.1.10) \qquad H_N \subset H_D.$$

We now state some basic results guaranteeing the existence of eigenvalues and eigenfunctions of (7.1.1)–(7.1.2) (cf. Gould [7.11]).

THEOREM 7.1. *With the assumptions of (7.1.6) and (7.1.7), assume that bounded sets in H_N are precompact in H_D. Then, the eigenvalue problem (7.1.1)–(7.1.2) has countably many real eigenvalues $0 < \lambda_1 \leq \lambda_2 \leq \cdots \leq \lambda_k \leq \lambda_{k+1} \leq \cdots$, having no finite limit point, and a corresponding sequence of eigenfunctions $\{f_j(x)\}_{j=1}^\infty$ with $f_j \in H_N$ for all $j \geq 1$, such that*

$$(7.1.11) \qquad \mathfrak{N} f_j(x) = \lambda_j \mathfrak{M} f_j(x), \qquad\qquad j = 1, 2, \cdots,$$

in Ω. These eigenfunctions can be chosen to be orthonormal in H_D, i.e.,

$$(7.1.12) \qquad (f_i, f_j)_D = \delta_{i,j} \quad \text{for all} \quad i, j \geq 1,$$

and $\{f_j\}_{j=1}^\infty$ is complete in H_D. Moreover, if

$$(7.1.13) \qquad R[w] \equiv \frac{\|w\|_N^2}{\|w\|_D^2} \quad \text{for all} \quad w \in H_N, \qquad\qquad w \neq 0,$$

denotes the Rayleigh-quotient, then

$$(7.1.14) \quad \begin{aligned} \lambda_k &= \min\{R[w] : w \in H_N, \ w \neq 0, \ (w, f_i)_D = 0, \ 1 \leq i \leq k - 1\} \\ &= R[f_k] \quad \textit{for all} \quad k \geq 1. \end{aligned}$$

7.2. The Rayleigh–Ritz method. Let S_M be any finite-dimensional subspace of H_N of dimension M. The Rayleigh–Ritz method for computing approximate eigenvalues and eigenfunctions of (7.1.1)–(7.1.2) consists of finding the extreme values and critical points of $R[w]$ over S_M, rather than all of H_N. If $\{w_i(x)\}_{i=1}^M$ is a basis for S_M, then the Rayleigh–Ritz method consists of finding the eigenvalues and eigenvectors of the associated matrix eigenvalue problem

$$(7.2.1) \quad A_M \mathbf{u} = \lambda B_M \mathbf{u}.$$

The $M \times M$ matrices $A_M \equiv (\alpha_{i,j}^M)$ and $B_M \equiv (\beta_{i,j}^M)$ have entries given explicitly by

$$(7.2.2) \quad \alpha_{i,j}^M \equiv (w_i, w_j)_N, \quad \beta_{i,j}^M \equiv (w_i, w_j)_D, \qquad 1 \leq i, j \leq M,$$

and, because of the assumptions of (7.1.7)–(7.1.9), the matrices A_M and B_M are real, symmetric, and positive definite. As such, (7.2.1) has M positive eigenvalues $0 < \hat{\lambda}_1 \leq \cdots \leq \hat{\lambda}_M$, the *approximate eigenvalues*, and M corresponding eigenvectors $\hat{\mathbf{u}}_1, \cdots, \hat{\mathbf{u}}_M$. Forming

$$(7.2.3) \quad \hat{f}_k(x) = \sum_{i=1}^M \hat{u}_{k,i} w_i(x), \qquad 1 \leq k \leq M,$$

where the $\hat{u}_{k,i}$ are the vector components of $\hat{\mathbf{u}}_k$, then $\{\hat{f}_k\}_{k=1}^M$ are the *approximate eigenfunctions*, associated with subspace S_M. The functions $\{\hat{f}_j(x)\}_{j=1}^M$ can be chosen to be orthonormal in H_D, i.e., in analogy with (7.1.12),

$$(7.2.4) \quad (\hat{f}_i, \hat{f}_j)_D = \delta_{i,j}, \qquad 1 \leq i, j \leq M,$$

and, moreover, the approximate eigenvalues $\hat{\lambda}_k$ satisfy the following analogue of (7.1.14):

$$(7.2.5) \quad \hat{\lambda}_k = \min\{R[w] : w \in S_M, \ w \neq 0, (w, \hat{f}_i)_D = 0, 1 \leq i \leq k - 1\}.$$

We now state a result, which follows effectively from Birkhoff, de Boor, Swartz and Wendroff [7.5] (see [7.6]).

THEOREM 7.2. *With the assumptions of (7.1.6)–(7.1.7) assume that bounded sets in H_N are precompact in H_D. If S_M is any M-dimensional subspace of H_N with $M \geq k$, and $\{\tilde{f}_j\}_{j=1}^k \subset S_M$ is any globally approximating set of functions to the first k eigenfunctions of (7.1.1)–(7.1.2) in the norm $\|\cdot\|_D$, i.e.,*

$$(7.2.6) \quad \sum_{i=1}^k \|f_i - \tilde{f}_i\|_D^2 < 1,$$

then

$$(7.2.7) \qquad \lambda_j \leq \hat{\lambda}_j \leq \lambda_j + \frac{\displaystyle\sum_{i=1}^{j} \| f_i - \tilde{f}_i \|_N^2}{\left\{ 1 - \sqrt{\displaystyle\sum_{i=1}^{j} \| f_i - \tilde{f}_i \|_D^2} \right\}^2} \qquad for\ all \quad 1 \leq j \leq k,$$

*where $\hat{\lambda}_j$ is the approximate eigenvalue corresponding to λ_j for the subspace S_M. If,
moreover, the first k eigenvalues of (7.1.1)–(7.1.2) satisfy $0 < \lambda_1 < \lambda_2 < \cdots < \lambda_k$,
then there exists a constant K, independent of the choice of S_M, such that*

$$(7.2.8) \qquad \| f_k - \hat{f}_k \|_N \leq K \left\{ \sum_{j=1}^{k} (\hat{\lambda}_j - \lambda_j) \right\}^{1/2}.$$

Next, let $\{ S_{M_t} \}_{t=1}^{\infty}$ be a given (not necessarily nested) sequence of finite-dimensional subspaces of H_N, with $\dim S_{M_t} = M_t \geq k$ for all $t \geq 1$. The Rayleigh–Ritz
method applied to S_{M_t} yields M_t approximate eigenvalues $\{ \hat{\lambda}_{k,t} \}_{k=1}^{M_t}$ and M_t
approximate eigenfunctions $\{ \hat{f}_{k,t}(x) \}_{k=1}^{M_t}$ which are orthonormal in H_D, i.e.,

$$(\hat{f}_{i,t}, \hat{f}_{j,t})_D = \delta_{i,j}, \qquad\qquad 1 \leq i,j \leq M_t.$$

As an immediate consequence of Theorem 7.2, we have the following corollary.

COROLLARY 7.3. *With the assumptions of (7.1.6)–(7.1.7), assume that bounded
sets in H_N are precompact in H_D, and that the first k eigenvalues of (7.1.1)–(7.1.2)
satisfy $0 < \lambda_1 < \lambda_2 < \cdots < \lambda_k$. If*

$$(7.2.9) \qquad \lim_{t \to \infty} \left\{ \inf_{w \in S_{M_t}} \| w - f_j \|_N \right\} = 0 \quad for\ each \quad 1 \leq j \leq k,$$

then $\hat{\lambda}_{j,t} \downarrow \lambda_j$ as $t \to \infty$ for $1 \leq j \leq k$, and $\| \hat{f}_{j,t} - f_j \|_N \to 0$ as $t \to \infty$ for $1 \leq j \leq k$.

We now apply the results of Theorem 7.2 to give more typical-looking error
estimates for the Rayleigh–Ritz approximate eigenvalues and approximate
eigenfunctions in a subspace S_M of H_N. First, we assume that S_M, a finite-dimensional subspace of H_N, depends on a *single* mesh parameter $h > 0$, i.e., $S_M = S^h$
$\subset H_N$. For $p \geq m$, assume that the following approximation-theoretic error
bound for S^h is valid, i.e., there exists a positive constant K such that for all $0 < h$
≤ 1,

$$(7.2.10) \qquad \inf_{w \in S^h} \| w - v \|_{W_2^m(\Omega)} \leq K h^{p+1-m} \| v \|_{W_2^{p+1}(\Omega)} \qquad for\ all \quad v \in H_N \cap W_2^{p+1}(\Omega).$$

Such approximations are, as we have seen, typical (cf. § 3.1, § 5.1 and § 6.2). Next, it
is useful to assume that the boundary conditions of (7.1.2) are such that there is a
constant K for which

$$(7.2.11) \qquad\qquad \| v \|_N \leq K \| v \|_{W_2^m(\Omega)} \quad for\ all \quad v \in H_N;$$

for one-dimensional problems ($n = 1$), this is of course no added assumption. Note that from the first inequality of (7.1.7), the above inequality of (7.2.11) implies that

$$(7.2.11') \qquad \|v\|_D \leq K\|v\|_{W_2^m(\Omega)} \quad \text{for all} \quad v \in H_N.$$

This brings us to the next theorem.

THEOREM 7.4. *With the assumptions of Theorem 7.2, (7.2.10) and (7.2.11), assume that for a positive integer k, each eigenfunction $f_j(x)$ of (7.1.1)–(7.1.2) is an element of $W_2^{p+1}(\Omega)$ for $1 \leq j \leq k$, where $p \geq m$. Then, for h sufficiently small, the approximate Rayleigh–Ritz eigenvalues $\hat{\lambda}_j^h$ for S^h satisfy*

$$(7.2.12) \qquad \lambda_j \leq \hat{\lambda}_j^h \leq \lambda_j + Kh^{2(p+1-m)}, \qquad\qquad 1 \leq j \leq k.$$

Similarly, if $0 < \lambda_1 < \cdots < \lambda_k$, then for h sufficiently small, the approximate Rayleigh–Ritz eigenfunctions $\hat{f}_j^h(x)$ for S^h satisfy

$$(7.2.13) \qquad \|f_j - \hat{f}_j^h\|_N \leq Kh^{p+1-m}, \qquad\qquad 1 \leq j \leq k.$$

Proof. With k fixed, the approximation assumption of (7.2.10), coupled with the inequality of (7.2.11') implies that there are elements $\tilde{f}_j^h$ in S^h for $1 \leq j \leq k$ for which

$$\|f_j - \tilde{f}_j^h\|_D \leq K\|f_j - \tilde{f}_j^h\|_{W_2^m(\Omega)} \leq Kh^{p+1-m}\|f_j\|_{W_2^{p+1}(\Omega)}.$$

Thus, as $p \geq m$, we see that for h sufficiently small, $\{\tilde{f}_j^h(x)\}_{j=1}^k$ is evidently a *globally approximating* set in S^h to $\{f_j(x)\}_{j=1}^k$ in the norm $\|\cdot\|_D$ (cf. (7.2.6)). In the same way, the inequalities of (7.2.10) and (7.2.11) combine to give

$$\|f_j - \tilde{f}_j^h\|_N \leq Kh^{p+1-m}\|f_j\|_{W_2^{p+1}(\Omega)}, \qquad\qquad 1 \leq j \leq k.$$

Thus, the above two inequalities when applied to the inequalities of (7.2.7) of Theorem 7.2, give the desired eigenvalue error bounds of (7.2.12). Then it is easy to see that the inequalities of (7.2.12), when applied to the inequalities of (7.2.8), give (7.2.13).

We remark that, under the assumptions of Theorem 7.4, the exponent of h in (7.2.12) for eigenvalue approximation is *sharp*, i.e., it cannot in general be increased. This sharpness follows from results of Birkhoff and de Boor [7.12]. The same is true for the approximate eigenfunction error bounds of (7.2.13), *in the norm* $\|\cdot\|_N$. In the next section, we shall see that improved approximate eigenfunction error bounds can be obtained in other norms.

Before concluding this section, it is worthwhile to comment on the approximation-theoretic assumption of (7.2.10) for the subspace S^h. The real catch is that S^h is to be a finite-dimensional subspace of H_N, which means that the boundary conditions associated with H_N (cf. (7.1.2)) must be fulfilled. For *general* bounded regions Ω of R^n, this is a difficult assignment. However, there are ways of avoiding this difficulty. One way is to assume that the boundary conditions of (7.1.2) are all, in a Rayleigh–Ritz setting, *suppressible*, i.e., no boundary conditions appear in the variational formulation. Examples of this would be the cases of *periodic* boundary conditions, as treated by Strang and Fix [7.13], or Neumann-type boundary conditions.

Another way to approach the approximation-theoretic assumption of (7.2.10) is to *restrict* the generality of the domain Ω in R^n. For the case when Ω is a rectangular parallelepiped in R^n, and when the particular homogeneous boundary conditions

$$(7.1.4') \qquad D^\beta u(x) = 0, \quad 0 \leqq |\beta| \leqq m - 1, \quad x \in \partial\Omega,$$

are chosen in (7.1.4), i.e., no mixing of boundary conditions on portions of $\partial\Omega$ is permitted, then the tensor products of one-dimensional spline functions can be chosen to satisfy both the essential boundary conditions of (7.1.4'), as well as the approximation-theoretic assumption of (7.2.10). This approach has been taken by Schultz [7.7].

7.3. Improved approximate eigenfunction error bounds. As in the previous section, let $\{S_{M_t}\}_{t=1}^\infty$ be a sequence of finite-dimensional subspaces of H_N with $\dim S_{M_t} = M_t \geqq k$ for all $t \geqq 1$, and let $\hat{\lambda}_{k,t}$ and $\hat{f}_{k,t}$ be the approximate eigenvalue and eigenfunction in S_{M_t}, corresponding to the eigenvalue λ_k and eigenfunction f_k of (7.1.1)–(7.1.2). Now, let $\bar{f}_{k,t}$ be the N-norm projection of f_k onto S_{M_t}, i.e.,

$$(7.3.1) \qquad (f_k - \bar{f}_{k,t}, w)_N = 0 \quad \text{for all} \quad w \in S_{M_t}.$$

Since S_{M_t} is a finite-dimensional subspace of the Hilbert space H_N, $\bar{f}_{k,t}$ exists and is unique for all $1 \leqq k \leqq M_t, t \geqq 1$. We next state a result from [7.8].

THEOREM 7.5. *With the assumptions of* (7.1.6.)–(7.1.7), *assume that bounded sets in H_N are precompact in H_D, and let $\{S_{M_t}\}_{t=1}^\infty$ be any sequence of finite-dimensional subspaces of H_N such that* (7.2.9) *is satisfied for all $1 \leqq j \leqq k$. Then, if λ_k is a simple eigenvalue of* (7.1.1)–(7.1.2), *there exists a positive integer t_j such that*

$$(7.3.2) \qquad \|\bar{f}_{k,t} - \hat{f}_{k,t}\|_N \leqq K \|f_k - \bar{f}_{k,t}\|_D \quad \text{for all} \quad t \geqq t_j.$$

We remark that a similar result holds without the assumption that λ_k is simple (cf. [7.8]). Continuing, it follows from the triangle inequality that

$$\|f_k - \hat{f}_{k,t}\|_D \leqq \|f_k - \bar{f}_{k,t}\|_D + \|\bar{f}_{k,t} - \hat{f}_{k,t}\|_D \leqq K \|f_k - \bar{f}_{k,t}\|_D,$$

the last inequality following from (7.3.2), since $\|w\|_D \leqq K \|w\|_N$ (cf. (7.1.7)). We state the above conclusion as follows.

COROLLARY 7.6. *With the hypotheses of Theorem 7.5, then*

$$(7.3.3) \qquad \|f_k - \hat{f}_{k,t}\|_D \leqq K \|f_k - \bar{f}_{k,t}\|_D \quad \text{for all} \quad t \geqq t_j.$$

To give an application of this corollary, consider the N-norm projection $\bar{f}_{k,t}$ of f_k on the subspace S_{M_t}, as defined in (7.3.1). From the definitions of (7.1.9) and (7.3.1), we see that $\bar{f}_{k,t}$ is the *Galerkin approximation* in S_{M_t} of the solution of the elliptic boundary value problem

$$(7.3.4) \qquad \mathfrak{N}u = \mathfrak{N}f_k,$$

subject to the boundary conditions of (7.1.2). As in the previous section, we now consider a finite-dimensional subspace S^h of H_N which depends on the positive parameter h where $0 < h \leq 1$, and we assume that the following typical Galerkin error bound is valid for S^h (cf. Theorem 5.4 of § 5.2):

$$(7.3.5) \qquad \| f_k - \bar{f}^h_k \|_{W^r_2(\Omega)} \leq K h^\sigma \| f_k \|_{W^{p+1}_2(\Omega)}, \qquad f_k \in H_N \cap W^{p+1}_2(\Omega),$$

where $\bar{f}^h_k$ is the Galerkin approximation in S^h to f_k, and where $\sigma \equiv \min(p + 1 - r; 2(p + 1 - m))$, where $p + 1 \geq m$. Next, if the boundary conditions of (7.1.2) allow us to deduce that

$$(7.3.6) \qquad \| v \|_D \leq K \| v \|_{W^r_2(\Omega)} \quad \text{for all} \quad v \in H_D,$$

thus sharpening the inequality of (7.2.11'), then the error bound of (7.3.3) of Corollary 7.6, when coupled with (7.3.5) and (7.3.6), becomes

$$(7.3.7) \qquad \| f_k - \hat{f}^h_k \|_D \leq K h^\sigma \| f_k \|_{W^{p+1}_2(\Omega)}, \qquad f_k \in H_N \cap W^{p+1}_2(\Omega),$$

where $\sigma = \min(p + 1 - r; 2(p + 1 - m))$, and where $\hat{f}^h_k$ is the approximate eigenfunction corresponding to f_k in S^h. In particular, if $p + 1 \geq 2m - r$, then the above error bound becomes

$$(7.3.7') \qquad \| f_k - \hat{f}^h_k \|_D \leq K h^{p+1-r} \| f_k \|_{W^{p+1}_2(\Omega)},$$

which improves the related error bound of (7.2.13), since $r < m$. Similar improved results can be obtained in the uniform norm [7.9], and are related to results of Nitsche [7.14], [7.15].

REFERENCES

[7.1] R. COURANT, *Variational methods for the solution of problems of equilibrium and vibrations,* Bull. Amer. Math. Soc., 49 (1943), pp. 1–23.

[7.2] L. V. KANTOROVICH AND V. I. KRYLOV, *Approximate Methods of Higher Analysis,* Interscience, New York, 1958.

[7.3] G. BIRKHOFF AND C. R. DE BOOR, *Piecewise polynomial interpolation and approximation,* Approximation of Functions, H. L. Garabedian, ed., Elsevier, New York, 1965, pp. 164–190.

[7.4] B. WENDROFF, *Bounds for eigenvalues of some differential operators by the Rayleigh–Ritz method,* Math. Comp., 19 (1965), pp. 218–224.

[7.5] G. BIRKHOFF, C. DE BOOR, B. SWARTZ AND B. WENDROFF, *Rayleigh–Ritz approximation by piecewise cubic polynomials,* SIAM J. Numer. Anal., 3 (1966), pp. 188–203.

[7.6] P. G. CIARLET, M. H. SCHULTZ AND R. S. VARGA, *Numerical methods of high-order accuracy for nonlinear boundary value problems. III. Eigenvalue problems,* Numer. Math., 12 (1968), pp. 120–133.

[7.7] M. H. SCHULTZ, *Multivariate spline functions and elliptic problems,* Approximations with Special Emphasis on Spline Functions, I. J. Schoenberg, ed., Academic Press, New York, 1969, pp. 279–347.

[7.8] J. G. PIERCE AND R. S. VARGA, *Higher order convergence results for the Rayleigh–Ritz method applied to eigenvalue problems. I. Estimates relating Rayleigh–Ritz and Galerkin approximations to eigenfunctions,* SIAM J. Numer. Anal., to appear.

[7.9] ———, *II. Improved error bounds for eigenfunctions,* to appear.

[7.10] G. BIRKHOFF AND G. FIX, *Accurate eigenvalue computations for elliptic problems,* Numerical Solution of Field Problems in Continuum Physics, vol. II, SIAM-AMS Proceedings, G. Birkhoff and R. S. Varga, eds., American Mathematical Society, Providence, R.I., 1970, pp. 111–151.

[7.11] S. H. Gould, *Variational Methods for Eigenvalue Problems*, University of Toronto Press, Toronto, 1966.

[7.12] G. Birkhoff and C. de Boor, *Piecewise polynomial interpolation and approximation*, Approximation of Functions, H. L. Garabedian, ed., Elsevier, Amsterdam, 1965, pp. 164–190.

[7.13] G. Strang and G. Fix, *An Analysis of the Finite Element Method*, Prentice-Hall, Englewood Cliffs, N.J., to appear.

[7.14] J. Nitsche, *Ein Kriterium für die Quasi-Optimalität des Ritzschen Verfahrens*, Numer. Math., 11 (1968), pp. 346–348.

[7.15] ———, *Verfahren von Ritz und Spline-Interpolation bei Sturm–Liouville–Rantwertproblemen*, Ibid., 13 (1969), pp. 260–265.

CHAPTER 8

Parabolic Problems

8.1. The semidiscrete Galerkin approximations. Let Ω be a bounded region in R^n, $n \geq 1$. We then consider the initial value problem

$$(8.1.1) \qquad \frac{\partial u(x, t)}{\partial t} + \mathscr{L}u(x, t) = f(x, t), \qquad\qquad t > 0, \quad x \in \Omega,$$

subject to the homogeneous boundary conditions

$$(8.1.2) \qquad D^\beta u(x, t) = 0, \quad x \in \partial\Omega, \quad \text{for all} \quad 0 \leq |\beta| \leq m - 1, \quad t > 0,$$

and subject to the initial condition

$$(8.1.3) \qquad\qquad u(x, 0) = u_0(x), \qquad\qquad\qquad x \in \Omega.$$

We assume that the linear operator $\mathscr{L}$ of (8.1.1) is of the form

$$(8.1.4) \qquad \mathscr{L}v(x) \equiv \sum_{|\alpha|, |\beta| \leq m} (-1)^{|\beta|} D^\beta \{q_{\alpha\beta}(x) D^\alpha v(x)\},$$

and that its associated bilinear form $B(u, v)$, defined by

$$(8.1.5) \qquad\qquad B(u, v) = \sum_{|\alpha|, |\beta| \leq m} \int_\Omega q_{\alpha\beta} D^\alpha u D^\beta v \, dx, \qquad\qquad u, v \in \mathring{W}_2^m(\Omega),$$

is $\mathring{W}_2^m(\Omega)$-*elliptic*, i.e., there exists a constant $\rho > 0$ such that

$$(8.1.6) \qquad\qquad B(v, v) \geq \rho \|v\|^2_{W_2^m(\Omega)} \quad \text{for all} \quad v \in \mathring{W}_2^m(\Omega),$$

where $\mathring{W}_2^m(\Omega)$ again denotes the completion of all functions v in $C^\infty(R^n)$ with compact support in Ω, with respect to the norm

$$\|v\|^2_{W_2^m(\Omega)} \equiv \sum_{|\alpha| \leq m} \|D^\alpha v\|^2_{L_2(\Omega)}.$$

If we assume all the coefficients $q_{\alpha\beta}(x)$ are bounded in $\bar\Omega$, then it is known (cf. Browder [8.1], [8.2], and Lions [8.3]) that there is a unique generalized solution $u(x, t)$ of (8.1.1)–(8.1.2) in $\mathring{W}_2^m(\Omega)$ for each $t > 0$, which is continuously differentiable with respect to t, i.e., $\partial u(x, t)/\partial t$ is in $\mathring{W}_2^m(\Omega)$ for each $t > 0$. In analogy with (5.2.5), this unique generalized solution $u(x, t)$ of (8.1.1)–(8.1.2) satisfies

$$(8.1.7) \qquad \left(\frac{\partial u(\cdot, t)}{\partial t}, v\right)_0 + B(u(\cdot, t), v) = (f(\cdot, t), v)_0$$

$$\text{for all} \quad v \in \mathring{W}_2^m(\Omega), \quad \text{for all} \quad t > 0,$$

59

and

$$(8.1.8) \qquad (u(\cdot,0),v)_0 = (u_0,v)_0 \quad \text{for all} \quad v \in \mathring{W}_2^m(\Omega),$$

where as usual $(u,v)_0 \equiv \int_\Omega uv \, dx$.

To define a *semidiscrete Galerkin approximation* of the solution of (8.1.1)–(8.1.3), consider any finite-dimensional subspace S_M of $\mathring{W}_2^m(\Omega)$, with basis $\{w_i(x)\}_{i=1}^M$, and form

$$w(x,t) = \sum_{i=1}^M c_i(t)w_i(x), \qquad\qquad t \geqq 0, \quad x \in \Omega.$$

In analogy with (8.1.7)–(8.1.8), we say that

$$\hat{w}(x,t) = \sum_{i=1}^M \hat{c}_i(t)w_i(t)$$

is the *semidiscrete Galerkin approximation* in S_M of the solution of (8.1.1)–(8.1.3) if

$$(8.1.9) \qquad \left(\frac{\partial\hat{w}(\cdot,t)}{\partial t},v\right)_0 + B(\hat{w}(\cdot,t),v) = (f(\cdot,t),v)_0$$

$$\text{for all} \quad v \in S_M, \quad \text{for all} \quad t > 0,$$

and

$$(8.1.10) \qquad (\hat{w}(\cdot,0),v)_0 = (u_0,v)_0 \quad \text{for all} \quad v \in S_M.$$

The same existence and uniqueness theory which is applied to (8.1.7)–(8.1.8) also guarantees existence and uniqueness of a continuously differentiable (with respect to t) solution of (8.1.9)–(8.1.10). Note that $\hat{w}(x,0)$ is from (8.1.10) just the best L_2-approximation of u_0 in S_M.

Equations (8.1.9)–(8.1.10) are equivalent to the following system of M ordinary differential equations:

$$(8.1.11) \qquad \mathscr{B}\frac{d\hat{\mathbf{c}}(t)}{dt} + \mathscr{A}\hat{\mathbf{c}}(t) = \mathbf{k}(t), \qquad\qquad t > 0,$$

and initial conditions

$$(8.1.12) \qquad \mathscr{B}\hat{\mathbf{c}}(0) = \mathbf{g},$$

where $\hat{\mathbf{c}}(t) \equiv (\hat{c}_1(t),\hat{c}_2(t),\cdots,\hat{c}_M(t))^T$, and where $\mathscr{B} \equiv (\beta_{i,j})$ and $\mathscr{A} = (\alpha_{i,j})$ are the $M \times M$ real symmetric and positive definite matrices, defined explicitly by

$$(8.1.13) \qquad \beta_{i,j} \equiv (w_i,w_j)_0, \quad \alpha_{i,j} \equiv B(w_i,w_j), \qquad 1 \leqq i,j \leqq M,$$

and

$$k_i(t) = (f(\cdot,t),w_i)_0, \quad g_i = (u_0,w_i)_0, \qquad 1 \leqq i \leqq M.$$

The solution $\hat{c}(t)$ of (8.1.11)–(8.1.12) can be expressed as

$$(8.1.14) \quad \hat{\mathbf{c}}(t) = \exp(-t\mathscr{B}^{-1}\mathscr{A})\mathscr{B}^{-1}\mathbf{g} + \int_0^t \exp[(t'-t)\mathscr{B}^{-1}\mathscr{A}]\mathscr{B}^{-1}\mathbf{k}(t')\,dt'$$

for all $t \geq 0$, and standard techniques can be used to solve (8.1.11)–(8.1.12). For $\mathbf{k}(t)$ continuous in t, it is clear from (8.1.11) that $\hat{w}(x, t) \equiv \sum_{i=1}^{M} \hat{c}_i(t) w_i(x)$ is continuously differentiable with respect to t.

In order now to estimate $\|u(\,\cdot\,, t) - \hat{w}(\,\cdot\,, t)\|_{L_2(\Omega)}$, define $\tilde{w}(x, t)$ for each *fixed* $t \geq 0$ as the Galerkin approximation in S_M of the *steady-state* elliptic problem:

$$(8.1.15) \qquad \begin{aligned} \mathscr{L} v(x) &= \mathscr{L} u(x, t), \quad x \in \Omega, \\ D^\beta v(x) &= 0, \quad x \in \partial\Omega, \quad 0 \leq |\beta| \leq m - 1. \end{aligned}$$

Equivalently, this implies for each fixed $t \geq 0$ that

$$(8.1.15') \qquad B(\tilde{w}(\,\cdot\,, t), v) = B(u(\,\cdot\,, t), v) \quad \text{for all} \quad v \in S_M \subset \mathring{W}_2^m(\Omega).$$

In other words, if $\tilde{w}(x, t) \equiv \sum_{i=1}^{M} \tilde{c}_i(t) w_i(t)$ and if $B(u(\,\cdot\,, t), w_i) \equiv h_i(t)$, then the coefficients $\tilde{c}_i(t)$ which determine $\tilde{w}(x, t)$ are given by the solution of the matrix problem $\mathscr{A} \tilde{\mathbf{c}}(t) = \mathbf{h}(t)$, where the nonsingular matrix $\mathscr{A} = (\alpha_{i,j})$ is defined in (8.1.13). Next, from (8.1.7), we can write

$$\left(\frac{\partial}{\partial t}[u(\,\cdot\,, t) - \tilde{w}(\,\cdot\,, t)], v \right)_0 = (f(\,\cdot\,, t), v)_0 - B(u(\,\cdot\,, t), v) - \left(\frac{\partial \tilde{w}(\,\cdot\,, t}{\partial t}, v \right)_0$$

$$\text{for all} \quad t > 0, \quad \text{for all} \quad v \in S_M,$$

and from (8.1.9) and (8.1.15'), the right-hand side becomes

$$\left(\frac{\partial \hat{w}(\,\cdot\,, t)}{\partial t}, v \right)_0 + B(\hat{w}(\,\cdot\,, t), v) - B(\tilde{w}(\,\cdot\,, t), v) - \left(\frac{\partial \tilde{w}(\,\cdot\,, t)}{\partial t}, v \right)_0.$$

Hence, combining these expressions and choosing $v = (\hat{w}(\,\cdot\,, t) - \tilde{w}(\,\cdot\,, t)) \in S_M$ for any fixed $t > 0$, gives simply

$$(8.1.16) \qquad \left(\frac{\partial}{\partial t}[u - \tilde{w}], \hat{w} - \tilde{w} \right)_0 = \left(\frac{\partial}{\partial t}[\hat{w} - \tilde{w}], \hat{w} - \tilde{w} \right)_0 + B(\hat{w} - \tilde{w}, \hat{w} - \tilde{w}),$$

where we have suppressed momentarily the time-dependence in the above expression. Using first Schwarz's inequality and then the elementary inequality $|cd| \leq (1/4\rho)c^2 + \rho d^2$, the term on the left in (8.1.16) can be bounded above by

$$\left(\frac{\partial}{\partial t}[u - \tilde{w}], \hat{w} - \tilde{w} \right)_0 \leq \left\| \frac{\partial}{\partial t}[u - \tilde{w}] \right\|_{L_2(\Omega)} \cdot \|\hat{w} - \tilde{w}\|_{L_2(\Omega)}$$

$$\leq \frac{1}{4\rho} \left\| \frac{\partial}{\partial t}[u - \tilde{w}] \right\|_{L_2(\Omega)}^2 + \rho \|\hat{w} - \tilde{w}\|_{L_2(\Omega)}^2.$$

Using the ellipticity assumption (8.1.6), the last term of (8.1.16) can be bounded below by $B(\hat{w} - \tilde{w}, \hat{w} - \tilde{w}) \geq \rho \|\hat{w} - \tilde{w}\|_{L_2(\Omega)}^2$. Combining these inequalities in (8.1.16) then gives

$$(8.1.17) \qquad \frac{d}{dt} \|\hat{w}(\,\cdot\,, t) - \tilde{w}(\,\cdot\,, t)\|_{L_2(\Omega)}^2 \leq \frac{1}{2\rho} \left\| \frac{\partial}{\partial t}[u(\,\cdot\,, t) - \tilde{w}(\,\cdot\,, t)] \right\|_{L_2(\Omega)}^2$$

since

$$\left(\frac{\partial \phi}{\partial t}, \phi\right)_0 \equiv \frac{1}{2}\frac{d}{dt}\|\phi\|^2_{L_2(\Omega)}.$$

Hence, integrating with respect to t gives

$$\|\hat{w}(\,\cdot\,, t) - \tilde{w}(\,\cdot\,, t)\|^2_{L_2(\Omega)} \leqq \|\hat{w}(\,\cdot\,, 0) - \tilde{w}(\,\cdot\,, 0)\|^2_{L_2(\Omega)}$$

$$+ \frac{T}{2\rho} \sup_{0 \leqq t \leqq T}\left\|\frac{\partial}{\partial t}[u(\,\cdot\,, t) - \tilde{w}(\,\cdot\,, t)]\right\|^2_{L_2(\Omega)}, \qquad 0 \leqq t \leqq T,$$

which we can write also as

$$(8.1.18) \qquad \|\hat{w}(\,\cdot\,, t) - \tilde{w}(\,\cdot\,, t)\|_{L_2(\Omega)} \leqq \|\hat{w}(\,\cdot\,, 0) - \tilde{w}(\,\cdot\,, 0)\|_{L_2(\Omega)}$$

$$+ \left(\frac{T}{2\rho}\right)^{1/2} \sup_{0 \leqq t \leqq T}\left\|\frac{\partial}{\partial t}u(\,\cdot\,, t) - \frac{\partial \tilde{w}}{\partial t}(\,\cdot\,, t)\right\|_{L_2(\Omega)}, \qquad 0 \leqq t \leqq T.$$

We now assume that the subspace S_M depends on a *single* mesh parameter $h > 0$, i.e., $S_M \equiv S^h \subset \mathring{W}^m_2(\Omega)$, such that the following error bound is valid:

$$(8.1.19) \quad \|v - \tilde{v}^h\|_{L_2(\Omega)} \leqq Kh^r\|v\|_{W^{p+1}_2(\Omega)} \quad \text{for all} \quad v \in W^{p+1}_2(\Omega) \cap \mathring{W}^m_2(\Omega),$$

where

$$(8.1.20) \qquad\qquad r \equiv \min(p + 1, 2(p + 1 - m)),$$

and where $p \geqq m$ and $\tilde{v}^h$ is the *Galerkin approximation* of v in S^h defined by $B(\tilde{v}^h, w) = B(v, w)$ for all $w \in S^h$, as in (8.1.15′). Note that the assumption of (8.1.19)–(8.1.20) is typical of the Galerkin error bounds for elliptic boundary value problems, as considered in Theorem 5.4 of § 5.2. With the assumption of (8.1.19), we can then prove the following theorem.

THEOREM 8.1. *Let $u(x, t)$ be the unique continuously differentiable (with respect to t) solution of (8.1.7)–(8.1.8) in $\mathring{W}^m_2(\Omega)$, and let $\hat{w}^h(x, t)$ be its semidiscrete Galerkin approximation in $S^h \subset \mathring{W}^m_2(\Omega)$ in the sense of (8.1.9)–(8.1.10), where S^h satisfies (8.1.19). Then, if $u(\,\cdot\,, t) \in W^{p+1}_2(\Omega)$ for each $t > 0$, where $p + 1 \geqq m$, then for some constant $K = K(T, u)$, and $r \equiv \min(p + 1, 2(p + 1 - m))$,*

$$(8.1.21) \qquad\qquad \|u(\,\cdot\,, t) - \hat{w}^h(\,\cdot\,, t)\|_{L_2(\Omega)} \leqq Kh^r \quad \text{for} \quad 0 \leqq t \leqq T.$$

Proof. From the triangle inequality, we surely have that

$$\|u(\,\cdot\,, t) - \hat{w}^h(\,\cdot\,, t)\|_{L_2(\Omega)} \leqq \|u(\,\cdot\,, t) - \tilde{w}^h(\,\cdot\,, t)\|_{L_2(\Omega)} + \|\tilde{w}^h(\,\cdot\,, t) - \hat{w}^h(\,\cdot\,, t)\|_{L_2(\Omega)}.$$

$$(8.1.22)$$

The first term on the right is suitably bounded above by the assumption of (8.1.19). Thus, it remains to bound the last term of (8.1.22), or from (8.1.18), the two terms on the right of (8.1.18).

First, since from (8.1.10), $\hat{w}^h(\,\cdot\,,0)$ is the best L_2-approximation of $u(\,\cdot\,,0)$ in S^h, then $\|\hat{w}^h(\,\cdot\,,0) - u(\,\cdot\,,0)\|_{L_2(\Omega)} \leqq \|\tilde{w}^h(\,\cdot\,,0) - u(\,\cdot\,,0)\|_{L_2(\Omega)}$. Consequently, from the triangle inequality and (8.1.19),

$$\|\hat{w}^h(\,\cdot\,,0) - \tilde{w}^h(\,\cdot\,,0)\|_{L_2(\Omega)} \leqq 2\|\tilde{w}^h(\,\cdot\,,0) - u(\,\cdot\,,0)\|_{L_2(\Omega)} \leqq Kh^r\|u(\,\cdot\,,0)\|_{W_2^{p+1}(\Omega)}.$$

This then bounds the first term on the right in (8.1.18). Next, it is important to note that the coefficients in the bilinear form for $B(u, v)$ in (8.1.5) are *time-independent*. Thus, differentiating with respect to t in (8.1.15') gives

$$B\left(\frac{\partial \tilde{w}^h(\,\cdot\,,t)}{\partial t}, v\right) = B\left(\frac{\partial u(\,\cdot\,,t)}{\partial t}, v\right) \quad \text{for all} \quad v \in S^h \subset \mathring{W}_2^m(\Omega).$$

This means that $\partial \tilde{w}^h(\,\cdot\,,t)/\partial t$ is the Galerkin approximation of $\partial u(\,\cdot\,,t)/\partial t$ in $\mathring{W}_2^m(\Omega)$. Hence, the error bound of (8.1.19) can be applied with $v = \partial u(\,\cdot\,,t)/\partial t$, i.e.,

$$\left\|\frac{\partial u(\,\cdot\,,t)}{\partial t} - \frac{\partial \tilde{w}^h(\,\cdot\,,t)}{\partial t}\right\|_{L_2(\Omega)} \leqq Kh^r\left\|\frac{\partial u(\,\cdot\,,t)}{\partial t}\right\|_{W_2^{p+1}(\Omega)},$$

which then bounds suitably the second term on the right in (8.1.18).

We first remark that the exponent r of h in (8.1.21), as given in (8.1.20), is *best possible*. This follows from discussions in § 5.2. Special cases of Theorem 8.1 were apparently first established in Price and Varga [8.4], but the discussion here parallels that of Strang and Fix [8.5] for the case $\Omega = R^n$. In addition, important contributions have been made by Douglas and Dupont [8.6], [8.7], Swartz and Wendroff [8.8], and Fix and Nassif [8.9], and these contributions will in part be discussed in § 8.3.

8.2. Stability considerations. We now examine (8.1.9) and (8.1.10). Setting $v = \hat{w}(\,\cdot\,,t)$ in (8.1.9), the fact that $B(\hat{w}, \hat{w})$ from (8.1.6) is *nonnegative* gives us that

$$(8.2.1) \qquad \frac{1}{2}\frac{d}{dt}\|\hat{w}(\,\cdot\,,t)\|^2_{L_2(\Omega)} \equiv \left(\frac{\partial \hat{w}(\,\cdot\,,t)}{\partial t}, \hat{w}(\,\cdot\,,t)\right)_0 \leqq (f(\,\cdot\,,t), \hat{w}(\,\cdot\,,t))_0$$

for all $t > 0$. Next, from Schwarz's inequality and from the elementary inequality $|ab| \leqq a^2/2 + b^2/2$, the last term of (8.2.1) can be bounded above by

$$(f(\,\cdot\,,t), \hat{w}(\,\cdot\,,t))_0 \leqq \tfrac{1}{2}\|\hat{w}(\,\cdot\,,t)\|^2_{L_2(\Omega)} + \tfrac{1}{2}\|f(\,\cdot\,,t)\|^2_{L_2(\Omega)}.$$

Thus, combining this inequality with the inequality of (8.2.1) gives

$$\frac{d}{dt}\|\hat{w}(\,\cdot\,,t)\|^2_{L_2(\Omega)} \leqq \|\hat{w}(\,\cdot\,,t)\|^2_{L_2(\Omega)} + \|f(\,\cdot\,,t)\|^2_{L_2(\Omega)}, \qquad\qquad t > 0,$$

$$(8.2.2)$$
$$\hat{w}(\,\cdot\,,0) \equiv \hat{v}.$$

As is easily verified, we can write the above expressions as

$$\frac{d}{dt}\{e^{-t}\|\hat{w}(\,\cdot\,,t)\|^2_{L_2(\Omega)}\} \leqq e^{-t}\|f(\,\cdot\,,t)\|^2_{L_2(\Omega)}, \qquad\qquad t > 0,$$

$$(8.2.2')$$
$$\hat{w}(\,\cdot\,,0) \equiv \hat{v},$$

and integrating these inequalities with respect to t then simply yields the following.

THEOREM 8.2. *The semidiscrete Galerkin approximation* $\hat{w}(x, t)$ *of the solution of* (8.1.1)–(8.1.3) *satisfies*

$$(8.2.3) \qquad \|\hat{w}(\cdot, t)\|^2_{L_2(\Omega)} \leq e^t \|\hat{v}\|^2_{L_2(\Omega)} + \int_0^t e^{(t-t')} \|f(\cdot, t')\|^2_{L_2(\Omega)}\, dt'$$

for all $t \geq 0$.

The importance of Theorem 8.2 is the following. For any fixed $T > 0$ and any fixed finite-dimensional subspace S_M of $\mathring{W}_2^m(\Omega)$, there exists a constant $K(T)$ such that

$$\|\hat{w}(\cdot, t)\|_{L_2(\Omega)} \leq K(T) \quad \text{for all} \quad 0 \leq t \leq T,$$

which implies the *uniform stability* in the L_2-norm of the semidiscrete approximation $\hat{w}(x, t)$.

With respect to the stability result of Theorem 8.2, it is clear that weaker hypotheses could be used even in the linear case to obtain a more general result of Theorem 8.2. For example, instead of (8.1.6), suppose that there exists a nonnegative constant such that

$$(8.2.4) \qquad B(u, u) \geq -\left(\frac{\rho}{2}\right) \|u\|^2_{L_2(\Omega)} \quad \text{for all} \quad u \in \mathring{W}_2^m(\Omega).$$

If we trace the steps leading to (8.2.3), the assumption of (8.2.4) similarly results in

$$(8.2.5) \quad \|\hat{w}(\cdot, t)\|^2_{L_2(\Omega)} \leq e^{(\rho+1)t} \|\hat{v}\|^2_{L_2(\Omega)} + \int_0^t e^{(\rho+1)(t-t')} \|f(\cdot, t')\|^2_{L_2(\Omega)}\, dt'.$$

This means that the uniform stability in the L_2-norm of the semidiscrete approximation $\hat{w}(x, t)$ holds rather generally. In particular, it covers the case of *dissipative operators* $\mathscr{L}$ of (8.1.4), i.e., operators $\mathscr{L}$ which satisfy

$$\mathrm{Re}\,(\mathscr{L}u, u)_0 \geq 0 \quad \text{for all} \quad u \in W_2^{2m}(\Omega).$$

Of course, for any *strongly elliptic operator* $\mathscr{L}$ of the form (8.1.4), it is well known that Gårding's inequality is satisfied, i.e., (cf. Yosida [8.10, p. 175]) there are positive constants k and K so that

$$(8.2.6) \qquad \|u\|^2_{W_2^m(\Omega)} \leq kB(u, u) + K\|u\|^2_{L_2(\Omega)} \quad \text{for all} \quad u \in \mathring{W}_2^m(\Omega),$$

and clearly, (8.2.6) implies (8.2.4).

8.3. Extensions. There are several ways in which one would want to generalize the result of Theorem 8.1. First, it seems rather reasonable to expect that the analysis just given could cover the case where the coefficients $q_{\alpha\beta}$ of $\mathscr{L}$ in (8.1.4) are mildly time-dependent. Even more desirable would be the study of the case where these coefficients $q_{\alpha\beta}$ were (mildly nonlinear) functions of the solution $u(x, t)$, particularly since the problems of petroleum reservoir mechanics are in fact of this type. Douglas and Dupont [8.7] have begun investigations of this type.

In [8.7], they consider the diffusion problem:

$$\frac{\partial u(x,t)}{\partial t} - \sum_{i,j=1}^{n} \frac{\partial}{\partial x_i}\left(a_{i,j}(x,u)\frac{\partial u}{\partial x_j}\right) = 0, \qquad x \in \Omega, \quad t > 0,$$

$$(8.3.1) \qquad u(x,t) = 0, \qquad\qquad\qquad x \in \partial\Omega, \quad t > 0,$$

$$u(x,0) = u_0(x), \qquad\qquad\qquad x \in \Omega,$$

where it is assumed that the $n \times n$ matrix $a_{i,j}(x,u)$ is real symmetric and uniformly positive definite, i.e., there exist positive constants η_1 and η_2 such that

$$(8.3.2) \qquad \eta_1 \sum_{i=1}^{n} \xi_i^2 \leq \sum_{i,j=1}^{n} a_{i,j}(x,s)\xi_i\xi_j \leq \eta_2 \sum_{i=1}^{n} \xi_i^2$$

for all $\xi \in R^n$, all $x \in \Omega$, and all $s \in R$. In addition, it is assumed that

$$(8.3.3) \qquad |a_{i,j}(x,r) - a_{i,j}(x,s)| \leqq K|r - s| \quad \text{for all} \quad x \in \Omega,$$

for any $-\infty < r, s < \infty$. Error bounds, somewhat weaker than those of Theorem 8.1, are derived for the corresponding semidiscrete Galerkin approximation. However, one of the more important features of [8.7] is the treatment of time-discretizations, such as the Crank–Nicolson method. If $\Delta t = T/N$, N a positive integer, then the *Crank–Nicolson–Galerkin approximation* $\hat{w}(\cdot, m\Delta t)$ of (8.3.1) is defined for a subspace $S_M \subset \mathring{W}_2^1(\Omega)$ as

$$(8.3.4) \qquad \left(\frac{\hat{w}_{m+1} - \hat{w}_m}{\Delta t}, v\right)_0 + B\left(\frac{\hat{w}_{m+1} + \hat{w}_m}{2}; \frac{\hat{w}_{m+1} + \hat{w}_m}{2}, v\right) = 0 \quad \text{for all} \quad v \in S_M,$$

$$(w_0, v)_0 = (u_0, v)_0 \quad \text{for all} \quad v \in S_M,$$

where $\hat{w}_m \equiv \hat{w}(\cdot, m\Delta t)$, and where

$$B(w;u,v) \equiv \int_\Omega \sum_{i,j=1}^{n} a_{i,j}(x,w(x))\frac{\partial u(x)}{\partial x_j} \cdot \frac{\partial v(x)}{\partial x_i}\, dx.$$

Under assumptions too lengthy to reproduce here, Douglas and Dupont show in [8.7] that the errors in the Crank–Nicolson–Galerkin method in an L_2-type norm for Hermite subspaces $H^{(m)}(\Omega)$, as described in § 3.1, with $\Omega \equiv (0,1) \times (0,1)$, is

$$(8.3.5) \qquad O(h^{2m-1} + (\Delta t)^2).$$

It appears doubtful that the exponent of h above is correct in an L_2-norm setting; certainly for coefficients $a_{i,j}$ in (8.3.1) which are independent of u, one knows that the correct exponent of h in (8.3.5) is $2m$.

One of the major drawbacks of the analysis of § 8.1 is the approximation assumption of (8.1.19); for *general* bounded regions Ω of R^n, it is extremely difficult to find such finite-dimensional subspaces S^h of $\mathring{W}_2^m(\Omega)$ which in fact do satisfy (8.1.19). This stems from the difficult assignment of having each element in S^h satisfy homogeneous Dirichlet data on $\partial\Omega$ (cf. (8.1.2)). But, there are at least three ways in which this "boundary problem" can be avoided. The first and second consist simply of having no boundary, in essence; the first considers the problem via Fourier analysis in *all* of R^n, as in § 5.2. But, as is easily seen, the analysis of § 8.1 is then adequate. The second method, which we shall describe in part below, consists of studying the problem (8.1.1) relative to *natural* or *Neumann* conditions on $\partial\Omega$. This has recently been considered by Fix and Nassif [8.9]. The third method, rather natural in light of the development in §§ 6.1–6.2, would be to consider *least squares* or penalty methods for such parabolic problems. This has very recently been studied by King [8.12].

As previously stated, suppose that the boundary conditions of (8.1.2) are replaced now by *natural* or *Neumann* boundary conditions. This has the effect of simply replacing the function space $\mathring{W}_2^m(\Omega)$ considered previously by $W_2^m(\Omega)$. With such boundary conditions, it is now easy from § 5.2 to find finite-dimensional subspaces S^h of $W_2^m(\Omega)$ for which the Galerkin approximation $\tilde{v}^h$ in S^h of v, i.e., $B(\tilde{v}^h, w) = B(v, w)$ for all $w \in S^h$, satisfies (cf. (8.1.19))

$$(8.3.6) \qquad \|v - \tilde{v}^h\|_{L_2(\Omega)} \leqq Kh^r \|v\|_{W_2^{p+1}(\Omega)},$$

where $r = \min(p + 1, 2(p + 1 - m))$. In studying such problems with natural boundary conditions, Fix and Nassif [8.9] have allowed time-dependent coefficients $q_{\alpha\beta}$ in (8.1.4). But the major result they obtain is *derivative* estimates of the error in the L_2-norm as well, something which (8.1.21) fails to do. Without giving any details, suffice it here to say that the error bounds of (8.1.21) of Theorem 8.1 are extended in [8.9] to the important cases

$$(8.3.7) \qquad \|u(\,\cdot\,, t) - \hat{w}^h(\,\cdot\,, t)\|_{W_2^j(\Omega)} \leqq Kh^{r-j}, \qquad 0 \leqq j \leqq r - 1.$$

The proofs make strong use of the notion of *quasi-interpolation* (cf. Theorem 5.3 and de Boor and Fix [8.11]).

REFERENCES

[8.1] F. E. Browder, *Non-linear equations of evolution*, Ann. of Math., 80 (1964), pp. 485–523.

[8.2] ———, *Non-linear initial value problems*, Ibid., 82 (1965), pp. 51–87.

[8.3] J. L. Lions, *Équationes Différentielles Opérationnelles et Problemes aux Limites*, Springer–Verlag, Berlin, 1961.

[8.4] H. S. Price and R. S. Varga, *Error bounds for semidiscrete Galerkin approximations of parabolic problems with applications to petroleum reservoir mechanics*, Numerical Solution of Field Problems in Continuum Physics, G. Birkhoff and R. S. Varga, eds., SIAM-AMS Proceedings, vol. 2, American Mathematical Society, Providence, R.I., 1970, pp. 74–94.

[8.5] G. Strang and G. Fix, *An Analysis of the Finite Element Method*, Prentice-Hall, Englewood Cliffs, N.J., to appear.

[8.6] J. Douglas and T. Dupont, *The numerical solution of waterflooding problems in petroleum engineering by variational methods*, Studies in Numerical Analysis 2, SIAM Publications, Philadelphia, 1970, pp. 53–63.

[8.7] ———, *Galerkin methods for parabolic equations*, SIAM J. Numer. Anal., 7 (1970), pp. 575–626.

[8.8] B. K. Swartz and B. Wendroff, *Generalized finite difference schemes*, Math. Comp., 23 (1969), pp. 37–49.

[8.9] G. Fix and N. Nassif, *Error bounds for derivatives and difference quotients for finite element approximation of parabolic problems*, Numer. Math., to appear.

[8.10] K. Yosida, *Functional Analysis*, Academic Press, New York, 1965.

[8.11] C. de Boor and G. Fix, *Spline approximation by quasi-interpolants*, J. Approx. Theory, to appear.

[8.12] J. T. King, *The approximate solution of parabolic initial boundary value problems by weighted least squares methods*, SIAM J. Numer. Anal., to appear.

CHAPTER 9

Chebyshev Semidiscrete Approximations for Linear Parabolic Problems

9.1. Introduction. The study of semidiscrete Galerkin techniques for parabolic partial differential equations in §8.1 was basically concerned with *spatial* approximations, as the time variable was left continuous in that treatment. The approximation theory used there emphasized approximation, by piecewise-polynomial functions, of elements in certain Sobolev spaces. The emphasis in this section rather is on *time* approximations, and this leads us to a completely different aspect of approximation theory, viz., the best Chebyshev *rational* approximation of e^{-x} (and reciprocals of certain entire functions) on $[0, +\infty)$.

To motivate our subsequent discussions, consider the solution $u(x, t)$ of the heat equation:

$$u_t(x, t) = u_{xx}(x, t) + r(x), \qquad 0 < x < 1, t > 0,$$

$$(9.1.1) \qquad u(x, 0) = \tilde{u}(x), \qquad 0 \leq x \leq 1,$$

$$u(0, t) = u(1, t) = 0, \qquad t > 0.$$

Leaving time continuous, consider the particular spatial discretization of (9.1.1) brought about by the usual three-point difference approximation to u_{xx}, i.e.,

$$u_{xx}(ih, t) \doteq \frac{u((i + 1)h, t) - 2u(ih, t) + u((i - 1)h, t)}{h^2}, \quad (N + 1)h = 1.$$

The resulting semidiscrete approximation $w(ih, t)$ to the solution $u(x, t)$ of (9.1.1) satisfies

$$\frac{dw(ih, t)}{dt} = \frac{w((i + 1)h, t) - 2w(ih, t) + w((i - 1)h, t)}{h^2} + r(ih),$$

$$1 \leq i \leq N, \quad t > 0,$$

$$(9.1.2) \qquad w(ih, 0) = \tilde{u}(ih), \qquad 0 \leq i \leq N + 1,$$

$$w(0, t) = w((N + 1)h, t) = 0, \qquad t > 0.$$

Written equivalently in matrix notation, this becomes

$$\frac{d\mathbf{w}(t)}{dt} = -A\mathbf{w}(t) + \mathbf{r}, \qquad t > 0,$$

$$(9.1.3)$$

$$\mathbf{w}(0) = \tilde{\mathbf{w}},$$

where $\mathbf{w}(t)$, $\mathbf{r}$, and $\tilde{\mathbf{w}}$ are column vectors with N components, with $\mathbf{w}(t) = (w_1(t), \cdots, w_N(t))^T$, where $w_i(t) \equiv w(ih, t)$. Note that $\mathbf{r}$ and $\tilde{\mathbf{w}}$ are determined from given quantities, and A is the familiar tridiagonal Hermitian and positive definite $N \times N$ matrix, given by

$$(9.1.4) \qquad A = \frac{1}{h^2} \begin{bmatrix} 2 & -1 & & & 0 \\ -1 & 2 & -1 & & \\ & \ddots & \ddots & \ddots & \\ & & & & -1 \\ 0 & & & -1 & 2 \end{bmatrix}.$$

In what is to follow, only the Hermitian positive definite character of the $N \times N$ matrix A is essential, *and we henceforth assume that our semidiscretization results in* (9.1.3) *with A Hermitian and positive definite*. In particular, this assumption is valid for linear parabolic problems in n-spatial variables of the form:

$$u_t(x, t) = \sum_{i=1}^{n} (K_i(x)u_{x_i}(x, t))_{x_i} - \sigma(x)u(x, t) + r(x),$$

$$t > 0, \quad x \in \Omega,$$

$$(9.1.5) \qquad u(x, 0) = \tilde{u}(x), \qquad\qquad\qquad\qquad x \in \Omega,$$

$$u(x, t) = g(x), \qquad\qquad\qquad x \in \partial\Omega, \quad t > 0,$$

where Ω is a bounded region in R^n, and the quantities $K_i(x)$, $\sigma(x)$ are positive in $\bar{\Omega}$, provided that a suitable $(2n + 1)$-point difference approximation of (9.1.5) is used (cf. [9.1, p. 253]).

Returning to (9.1.3), the solution $\mathbf{w}(t)$ can obviously be expressed as

$$(9.1.6) \qquad \mathbf{w}(t) = A^{-1}\mathbf{r} + \exp(-tA)\{\tilde{\mathbf{w}} - A^{-1}\mathbf{r}\}, \qquad t \geq 0,$$

where as usual, $\exp(-tA) \equiv \sum_{k=0}^{\infty}(-tA)^k/k!$. The solution of (9.1.6) is commonly approximated by means of matrix Padé rational approximations of $\exp(-tA)$, and these give, as special cases, the well-known forward difference, backward difference, and Crank–Nicolson methods for such parabolic problems (cf. [9.1, § 9.3]). Our interest in the next section will be on *Chebyshev*, rather than *Padé*, rational approximations of $\exp(-tA)$. This is because Padé rational approximations of e^{-x}, being defined as *local* approximations of e^{-x} at $x = 0$, are generally poor approximations of e^{-x} for large x, and this leads to restrictions (for reasons of stability and/or accuracy) on the time step that can be taken. Chebyshev rational approximations of e^{-x}, in contrast, are defined *globally* with respect to the interval $[0, +\infty)$, and do not have such time step restrictions, as we shall see.

9.2. Chebyshev semidiscrete approximations. To define the Chebyshev semidiscrete approximations of (9.1.6), we consider the following approximation problem. If π_m denotes all real polynomials $p(x)$ of degree at most m, and $\pi_{m,n}$

analogously denotes all real rational functions $r_{m,n}(x) = p(x)/q(x)$ with $p \in \pi_m$, $q \in \pi_n$, then let

$$(9.2.1) \qquad \lambda_{m,n} \equiv \inf_{\pi_{m,n}} \|e^{-x} - r_{m,n}(x)\|_{L_\infty[0,\infty]} = \inf_{\pi_{m,n}} \left\{ \sup_{x \geq 0} |e^{-x} - r_{m,n}(x)| \right\}.$$

These constants $\lambda_{m,n}$ are called the *Chebyshev constants for* e^{-x} with respect to the interval $[0, +\infty)$. It is obvious that $\lambda_{m,n}$ is finite if and only if $0 \leq m \leq n$, and moreover, given any pair (m, n) of nonnegative integers with $0 \leq m \leq n$, it is known (cf. Achieser [9.2, p. 55]) that, after dividing out possible common factors, there exists a unique $\hat{r}_{m,n} \in \pi_{m,n}$ with

$$(9.2.2) \qquad \hat{r}_{m,n}(x) = \hat{p}_{m,n}(x)/\hat{q}_{m,n}(x)$$

and with $\hat{q}_{m,n}(x) > 0$ on $[0, \infty)$, such that

$$(9.2.3) \qquad \lambda_{m,n} = \|e^{-x} - \hat{r}_{m,n}(x)\|_{L_\infty[0,\infty]}.$$

Since $\hat{q}_{m,n}(tA) = \sum_{j=0}^{n} c_j \cdot (tA)^j$ is a real polynomial in the $N \times N$ matrix A, it is evident from the fact that $\hat{q}_{m,n}(x)$ is positive on $[0, +\infty)$ that $\hat{q}_{m,n}(tA)$ is a Hermitian and positive definite $N \times N$ matrix for each $t \geq 0$. Thus, in analogy with (9.1.6), we define the (m, n)th *Chebyshev semidiscrete approximation* $w_{m,n}(t)$ of the solution $\mathbf{w}(t)$ of (9.1.3) as

$$(9.2.4) \qquad \mathbf{w}_{m,n}(t) = A^{-1}\mathbf{r} + (\hat{q}_{m,n}(tA))^{-1}(\hat{p}_{m,n}(tA))\{\tilde{\mathbf{w}} - A^{-1}\mathbf{r}\}, \qquad t \geq 0.$$

For the practical computation of $\mathbf{w}_{m,n}(t)$ for a fixed finite $t \geq 0$, assume first that the steady-state solution $\hat{\mathbf{w}} \equiv A^{-1}\mathbf{r}$ of (9.1.3) has been determined, which amounts to solving the matrix equation $A\hat{\mathbf{w}} = \mathbf{r}$. Then, we write (9.2.4) equivalently as

$$(9.2.5) \qquad \hat{q}_{m,n}(tA) \cdot \mathbf{w}_{m,n}(t) = \mathbf{v}_0, \quad \mathbf{v}_0 \equiv \hat{q}_{m,n}(tA)\hat{\mathbf{w}} + \hat{p}_{m,n}(tA)\{\tilde{\mathbf{w}} - \hat{\mathbf{w}}\},$$

where $\mathbf{v}_0$ is determined from the known initial vector $\tilde{\mathbf{w}}$ (cf. (9.1.3)), and the known steady-state vector $\hat{\mathbf{w}} = A^{-1}\mathbf{r}$. Since $\hat{q}_{m,n} \in \pi_n$ is positive on $[0, +\infty)$, $\hat{q}_{m,n}$ can be factored into real linear and quadratic factors:

$$(9.2.6) \qquad \hat{q}_{m,n}(x) = \prod_{i=1}^{s_1} l_i(x) \cdot \prod_{j=1}^{s_2} m_j(x), \quad s_1 + 2s_2 = n,$$

where $l_i \in \pi_1$, $m_j \in \pi_2$, and where the l_i and m_j are also positive on $[0, +\infty)$. Thus, the matrices $l_i(tA)$ and $m_j(tA)$ are again Hermitian and positive definite for each $t \geq 0$, and the solution $\mathbf{w}_{m,n}(t)$ of (9.2.5) can be obtained by solving recursively the matrix problems:

$$(9.2.7) \qquad \begin{aligned} m_j(tA)\mathbf{v}_j &= \mathbf{v}_{j-1}, & 1 \leq j \leq s_2, \\ l_i(tA)\mathbf{v}_{s_2+i} &= \mathbf{v}_{s_2+i-1}, & 1 \leq i \leq s_1, \end{aligned}$$

and then defining $\mathbf{w}_{m,n}(t) \equiv \mathbf{v}_{s_2+s_1}$. In particular, when A is tridiagonal as in (9.1.4), the matrices of (9.2.7) are either tridiagonal or five-diagonal positive definite matrices. As such, the solution of (9.2.7) by means of Gaussian elimination with no pivoting is both computationally fast and numerically accurate.

For computational efficiency, one should always choose $m = n$ in (9.2.4) for applications of the Chebyshev semidiscrete method to actual problems. The reason for this is quite clear: the bulk of the work in finding the solution $\mathbf{w}_{m,n}(t)$ of (9.2.5) comes from the inversion of the polynomial $\hat{q}_{m,n}(tA)$ of degree n in the matrix A, and the work involved in this inversion in practice is virtually independent of the choice of m. For further discussion of such computational aspects of the Chebyshev semidiscrete method, see [9.3].

To estimate the error in $\mathbf{w}(t) - \mathbf{w}_{m,n}(t)$, we use vector l_2-norms, i.e., if $\mathbf{v} = (v_1, \cdots, v_N)^T$, then $\|\mathbf{v}\|_2^2 \equiv \sum_{i=1}^{N} |v_i|^2$. If, for any $N \times N$ matrix C, $\|C\|_2$ denotes the induced operator norm (or spectral norm) of C, i.e.,

$$(9.2.8) \qquad \|C\|_2 \equiv \sup_{\mathbf{v} \neq \mathbf{0}} \left\{ \frac{\|C\mathbf{v}\|_2}{\|\mathbf{v}\|_2} \right\},$$

it is well known (cf. [9.1, p. 11]) when C is Hermitian with (real) eigenvalues μ_i, $1 \leq i \leq N$, that $\|C\|_2$ can be expressed as

$$(9.2.9) \qquad \|C\|_2 = \max_{1 \leq i \leq N} |\mu_i|.$$

Consequently, if $\{\lambda_i\}_{i=1}^{N}$ denotes the (positive) eigenvalues of A, the assumed Hermitian character of A allows us to conclude from (9.2.9) that

$$(9.2.10) \qquad \|\exp(-tA) - \hat{r}_{m,n}(tA)\|_2 = \max_{1 \leq i \leq N} |e^{-t\lambda_i} - \hat{r}_{m,n}(t\lambda_i)|$$

$$\text{for all} \quad t \geq 0.$$

But as $t\lambda_i \geq 0$ for all $1 \leq i \leq N$ and for all $t \geq 0$, it follows from (9.2.3) that

$$\|\exp(-tA) - \hat{r}_{m,n}(tA)\|_2 \leq \lambda_{m,n} \quad \text{for all} \quad t \geq 0.$$

Consequently, from (9.1.6) and (9.2.4),

$$(9.2.11) \qquad \|\mathbf{w}(t) - \mathbf{w}_{m,n}(t)\|_2 \leq \|\exp(-tA) - \hat{r}_{m,n}(tA)\|_2 \cdot \|\tilde{\mathbf{w}} - A^{-1}\mathbf{r}\|_2$$

$$\leq \lambda_{m,n}\|\tilde{\mathbf{w}} - A^{-1}\mathbf{r}\|_2 \quad \text{for all} \quad t \geq 0.$$

Note that since the right-hand side of (9.2.11) is *independent* of t, we have an error bound for $\mathbf{w}(t) - \mathbf{w}_{m,n}(t)$ for *all* $t \geq 0$. In contrast with the familiar Padé methods which restrict the size of t for reasons of accuracy and/or stability, the Chebyshev semidiscrete method can be used for very large values of t. The difference, of course, comes from the fact that Padé rational approximations of e^{-x} are designed to approximate e^{-x} well in a neighborhood of $x = 0$, whereas Chebyshev rational approximations of e^{-x} are designed to approximate e^{-x} over $[0, +\infty)$.

In general, the error of the spatial discretization leading to (9.1.3) must be bounded to give the total error (i.e., space and time) of these Chebyshev semi-discrete approximations. Such spatial discretization errors have been discussed in §8.1.

9.3. The Chebyshev constants for e^{-x}. The utility of the Chebyshev semidiscrete approximations depends, from (9.2.11), on the behavior of the Chebyshev constants $\lambda_{m,n}$ of (9.2.1), as $n \to \infty$. From (9.2.1) it is clear that

$$(9.3.1) \qquad 0 < \lambda_{n,n} \leq \lambda_{n-1,n} \leq \cdots \leq \lambda_{0,n}, \qquad\qquad n \geq 0.$$

Based on elementary arguments, the following result was proved in Cody, Meinardus and Varga [9.4].

THEOREM 9.1. *Let $\{m(n)\}_{n=0}^{\infty}$ be any sequence of nonnegative integers with $0 \leq m(n) \leq n$ for each $n \geq 0$. Then*

$$(9.3.2) \qquad \lim_{n \to \infty} \sup(\lambda_{m(n),n})^{1/n} \leq \frac{e^{-\alpha}}{2} < \tfrac{1}{2},$$

where $\alpha = 0.13923\ldots$ is the real solution of $2\alpha e^{2\alpha+1} = 1$. Moreover,

$$(9.3.3) \qquad \lim_{n \to \infty} \sup(\lambda_{0,n})^{1/n} \geq \tfrac{1}{6}.$$

The results of (9.3.2) and (9.3.3) establish the *geometric convergence to zero* of the Chebyshev constants $\lambda_{m,n}$ for e^{-x} in $[0, \infty)$. In particular, if $m(n) = n$, then the Chebyshev constants $\lambda_{n,n}$ for e^{-x} in $[0, +\infty)$ are from [9.4]:

n	$\lambda_{n,n}$	n	$\lambda_{n,n}$	n	$\lambda_{n,n}$
0	$5.00(-01)$	5	$9.35(-06)$	10	$1.36(-10)$
1	$6.69(-02)$	6	$1.01(-06)$	11	$1.47(-11)$
2	$7.36(-03)$	7	$1.09(-07)$	12	$1.58(-12)$
3	$7.99(-04)$	8	$1.17(-08)$	13	$1.70(-13)$
4	$8.65(-05)$	9	$1.26(-09)$	14	$1.83(-14)$

where $\alpha(-\beta)$ denotes $\alpha \cdot 10^{-\beta}$ in the table above. Thus, the rate of convergence to zero of the $\lambda_{n,n}$ appears to be much better than that given by the upper bound of (9.3.2). Also, the quantities $\lambda_{0,n}$, $0 \leq n \leq 9$, as tabulated in [9.4], would lead one to conjecture that $\lim_{n \to \infty} (\lambda_{0,n})^{1/n}$ exists, and that

$$(9.3.4) \qquad \lim_{n \to \infty} (\lambda_{0,n})^{1/n} = \tfrac{1}{3}.$$

This has in fact been recently shown by Schönhage [9.10].

9.4. Chebyshev constants for other entire functions. The preceding results on the geometric convergence to zero of the Chebyshev constants $\lambda_{m,n}$ for $1/e^x$ in (9.3.2) and (9.3.3) hold for a wider class of entire functions than just $f(z) = e^z$. A generalization of the results of Theorem 9.1 has been recently given in Meinardus and Varga [9.5], and can be described as follows.

Let $f(z) = \sum_{k=0}^{\infty} a_k z^k$ be an entire function (i.e., analytic for every finite z) with

$M_f(r) \equiv \sup_{|z|=r} |f(z)|$ its maximum modulus function. Then, f is of *perfectly regular growth* (ρ, B) (cf. Boas [9.6, p. 8] and Valiron [9.7, p. 45]) if there exist two (finite) positive numbers ρ (the order) and B (the type) such that

$$(9.4.1) \qquad\qquad \lim_{r \to \infty} \frac{\log M_f(r)}{r^\rho} = B.$$

We then have the following theorem (cf. [9.5]).

THEOREM 9.2. *Let* $f(z) = \sum_{k=0}^{\infty} a_k z^k$ *be an entire function of perfectly regular growth* (ρ, B) *with* $a_k \geq 0$ *for all* $k \geq 0$, *and for any pair* (m, n) *of nonnegative integers with* $0 \leq m \leq n$, *let*

$$(9.4.2) \qquad\qquad \lambda_{m,n} \equiv \inf_{\pi_{m,n}} \left\| \frac{1}{f(x)} - r_{m,n}(x) \right\|_{L_\infty[0,\infty]}$$

be its associated Chebyshev constants. Then, for any sequence $\{m(n)\}_{n=0}^{\infty}$ *of nonnegative integers with* $0 \leq m(n) \leq n$ *for each* $n \geq 0$,

$$(9.4.3) \qquad\qquad \limsup_{n \to \infty} (\lambda_{m(n),n})^{1/n} \leq 2^{-1/\rho} < 1.$$

Moreover,

$$(9.4.4) \qquad\qquad \limsup_{n \to \infty} (\lambda_{0,n})^{1/n} \geq 2^{-2-1/\rho}.$$

As special cases of Theorem 9.2, we have of course $f(z) = e^z$, $f(z) = \sinh(z^p)$ and $f(z) = J_p(iz)$ for p a nonnegative integer, where J_p denotes the Bessel function of the first kind. For $f(z) = e^z$, for which $\rho = B = 1$ in (9.4.1), the results of (9.4.3) and (9.4.4) are slightly weaker than those of (9.3.2) and (9.3.3) of Theorem 9.1.

The proofs of Theorems 9.1 and 9.2 depend upon estimating

$$\frac{1}{s_n(x)} - \frac{1}{f(x)},$$

where $s_n(z) = \sum_{k=0}^{n} a_k z^k$ is the nth partial sum of $f(z)$. It is shown in [9.5] that, under the hypotheses of Theorem 9.2,

$$\lim_{n \to \infty} \left(\left\| \frac{1}{s_n} - \frac{1}{f} \right\|_{L_\infty[0,\infty]} \right)^{1/n} = 2^{-1/\rho},$$

so that the upper bound of (9.4.3) cannot be improved using this specific technique.

Upon examining Theorem 9.2, we see that the bounds of (9.4.3) and (9.4.4) depend upon ρ, but not on B, and this suggests the possibility of extensions of Theorem 9.2 to entire functions which are of finite order, but not of perfectly regular growth. Such extensions have been considered in Meinardus, Reddy, Taylor and Varga [9.9], and we state a representative result which generalizes Theorem 9.2. For notation, let $\mathscr{E}(r, s)$, for given $r > 0$ and $s > 1$, denote the unique open ellipse in the complex plane with foci at $x = 0$ and $x = r$ and semimajor and semiminor axes a and b such that $b/a = (s^2 - 1)/(s^2 + 1)$. If $f(z)$ is any entire function, we set

$$(9.4.5) \qquad \tilde{M}_f(r, s) = \sup\{|f(z)| : z \in \mathscr{E}(r, s)\}.$$

THEOREM 9.3. *Let* $f(z) = \sum_{k=0}^{\infty} a_k z^k$ *be an entire function with nonnegative Taylor coefficients and* $a_0 > 0$. *If there exists real numbers* $s > 1$, $A > 0$, $\theta > 0$ *and* $r_0 > 0$ *such that*

$$(9.4.6) \qquad \tilde{M}_f(r, s) \leq A(\|f\|_{L_\infty[0,r]})^\theta \quad \text{for all} \quad r \geq r_0,$$

then there exists a real number $q \geq s^{1/(1+\theta)} > 1$ *and a sequence of real polynomials* $\{p_n(x)\}_{n=0}^{\infty}$ *with* $p_n \in \pi_n$ *for each* $n \geq 0$ *such that*

$$(9.4.7) \qquad \limsup_{n \to \infty} \left\{ \left\| \frac{1}{f(x)} - \frac{1}{p_n(x)} \right\|_{L_\infty[0,\infty]} \right\}^{1/n} = \frac{1}{q} < 1.$$

Note that (9.4.7) implies the geometric convergence to zero of the Chebyshev constants $\{\lambda_{m(n),n}\}_{n=0}^{\infty}$ of $1/f$ when $0 \leq m(n) \leq n$.

To motivate the next result, it is convenient to recall some classical results of Bernstein for polynomial approximation on *finite* intervals. Given a real-valued function $f \in C^0[-1, +1]$, let

$$(9.4.8) \qquad E_n(f) \equiv \inf_{\pi_n} \|f - p_n\|_{L_\infty[-1,+1]}.$$

If f is the restriction to $[-1, +1]$ of a function analytic in an ellipse in the complex plane with foci -1 and $+1$, then Bernstein proved (cf. Meinardus [9.8, p. 91]) that there exists a real number $q > 1$ such that

$$(9.4.9) \qquad \limsup_{n \to \infty} E_n^{1/n}(f) = \frac{1}{q} < 1.$$

Conversely, if (9.4.9) holds, Bernstein proved the *inverse* result (cf. Meinardus [9.8, p. 92]) that f is necessarily the restriction to $[-1, +1]$ of a function analytic in an ellipse in the complex plane with foci at -1 and $+1$. Consider then the results of Theorems 9.2 and 9.3. These give *sufficient* conditions on the entire function $f(z)$ so that the Chebyshev constants $\lambda_{m,n}$ of $1/f$, for $0 \leq m \leq n$, converge geometrically to zero as $n \to \infty$. In the spirit of Bernstein's classical inverse theorems, the following result of [9.9] gives *necessary* conditions for this geometric convergence.

THEOREM 9.4. *Let* $f(x) > 0$ *be a real continuous function on* $[0, \infty)$, *such that there exist a sequence of real polynomials* $\{p_n(x)\}_{n=0}^{\infty}$ *with* $p_n \in \pi_n$ *for all* $n \geq 0$, *and a real number* $q > 1$ *such that*

$$(9.4.10) \qquad \limsup_{n \to \infty} \left(\left\| \frac{1}{p_n} - \frac{1}{f} \right\|_{L_\infty[0,\infty]} \right)^{1/n} = \frac{1}{q} < 1.$$

Then, there exists an entire function $F(z)$ with $F(x) = f(x)$ for all $x \geqq 0$. Moreover, F is of finite order, i.e.,

$$\limsup_{r \to \infty} \frac{\log \log M_F(r)}{\log r} = \rho < \infty.$$

In addition, for each $s > 1$, there exist real numbers $K = K(q, s) > 0, \theta = \theta(q, s) > 1$ and $r_0 = r_0(q, s) > 0$ such that

$$(9.4.11) \qquad \tilde{M}_F(r, s) \leqq K(\| f \|_{L_\infty[0,r]})^\theta \quad \textit{for all} \quad r \geqq r_0.$$

Finally, to complement the preceding results of this section, it is shown in [9.9] that there exist entire functions $f(z)$ of finite order which are positive on $[0, +\infty)$ for which the Chebyshev constants $\lambda_{m,n}$ of $1/f$, for $0 \leqq m \leqq n$, *cannot* converge geometrically to zero as $n \to \infty$.

REFERENCES

[9.1] RICHARD S. VARGA, *Matrix Iterative Analysis*, Prentice-Hall, Englewood Cliffs, N.J., 1962.

[9.2] N. I. ACHIESER, *Theory of Approximation*, Frederick Ungar, New York, 1956.

[9.3] RICHARD S. VARGA, *Some results in approximation theory with applications to numerical analysis*, Numerical Solution of Partial Differential Equations, II, B. E. Hubbard, ed., Academic Press, New York, 1971, pp. 623–649.

[9.4] W. J. CODY, G. MEINARDUS AND R. S. VARGA, *Chebyshev rational approximations to e^{-x} in $[0, +\infty]$ and applications to heat-conduction problems*, J. Approx. Theory, 2 (1969), pp. 50–65.

[9.5] GÜNTER MEINARDUS AND RICHARD S. VARGA, *Chebyshev rational approximations to certain entire functions in $[0, +\infty)$*, Ibid., 3 (1970), pp. 300–309.

[9.6] RALPH P. BOAS, *Entire Functions*, Academic Press, New York, 1954.

[9.7] GEORGES VALIRON, *Lectures on the General Theory of Integral Functions*, Chelsea, New York, 1949.

[9.8] GÜNTER MEINARDUS, *Approximation of Functions: Theory and Numerical Methods*, Springer-Verlag, New York, 1967.

[9.9] G. MEINARDUS, A. R. REDDY, G. D. TAYLOR AND R. S. VARGA, *Converse theorems and extensions in Chebyshev rational approximation to certain entire functions in $[0, +\infty)$*, Bull. Amer. Math. So., 77 (1971), pp. 460–461.

[9.10] A. SCHÖNHAGE, *Zur rational Approximierbarkeit von e^{-x} über $[0, \infty)$*, J. Approx. Theory, to appear.